Swarming Biology and Control

Swarming Biology and Control

Wally Shaw, BSc

BeeCraft

© 2021 Martin W Shaw

First published in 2021 by Bee Craft Ltd.

All rights reserved. No part of this publication may be reproduced, stored in a retrieval system or transferred in any form or by any means, electronic, mechanical, photocopying, recording or otherwise, without the prior written permission of the publisher.

A catalogue record for this book is available from the British Library.

ISBN: 978-0-900147-10-4

Published in Great Britain by

Bee Craft Limited
1 Fox Covert
Fetcham
Surrey
KT22 9XD

Typeset by Buzzwords Editorial Ltd, Little Addington, Kettering, Northamptonshire NN14 4AX

Printed in Great Britain by printondemand-worldwide

ACKNOWLEDGEMENTS

All photographs and diagrams by Wally Shaw apart from pp 53b, 152b, 168 by Adrian Waring and pp 30, 46, 63, 69a–e, 71, 81, 84, 86, 118, 139, 161, 162, 163a, 164, 169, 173b, 210 by Claire Waring. Cartoons by Ellie Shaw

Dedication

To Jenny Shaw, my partner in life and beekeeping, best friend and sternest critic

CONTENTS

Foreword	xv
Introduction	xix

Part 1: The Biology of Swarming

1.1 Our experience of swarming and swarm control (a voyage of discovery) — 3

The Pagden method	4
The truth begins to dawn	6
Louis Snelgrove	6
Split boards	8
The Gerstung theory	9
Snelgrove's Method I	10
Snelgrove's Method II	11
Continuing failure	15
A Eureka moment	15
Method II (modified) – total success	16
The Snelgrove board	18
Swarming prevention and control	18
Summary	19

1.2 The reproductive strategy of the honey bee — 23

Sex alleles	24
Haploid/diploid drones	24
Drone production	25
Eusocial insects	26
Queen longevity	27
Contrasting reproductive strategies	27
Plant extremes	28
The Feminine Monarchie	*28*
Maximising genetic diversity	29
Potential loss of genetic diversity	31
Gene cross-overs	31
Evolution of queen longevity	33

	Queen substance	34
	Laying workers	35
	Worker police	35
	Queen mating	36
	The Cape honey bee	37
	The advantages of polyandry	40
	The downside of polyandry (apart from the risk of queen loss)	40
	A two-queen colony	41
	Kinship	42
	Genetic relationships	42
1.3	**How we reached our current understanding of swarming**	**45**
	Unreliable source of information	46
	More reliable knowledge	47
	Practical experience	47
	Working hypotheses	48
1.4	**The three types of queen cell – their origin and function**	**51**
	Reasons for queen cells being present in a hive	51
	Different types of queen cells	52
	Inferior (scrub) queens	57
	Ambiguous situations	58
	When emergency cells are significant	59
	Understanding the bees' programme	60
1.5	**Triggers for swarming and the start of queen cells**	**61**
	How does a honey bee colony know when to swarm?	61
	Preferred cavity size	62
	Second swarming period	63
	Adjustable space	64
	External conditions	64
1.6	**The start of queen cells to the issue of the prime swarm**	**67**
	At what point does the intention to swarm start?	68
	Laying workers?	70
	Colony unanimity	71

CONTENTS

	'Swarm control' bees	72
	Decisions – how are they made and who makes them?	73
	Positive decision	74
	Small numbers of queen cells	74
	The point of no return	75
	Use of the vibration signal	75
	Maximising resources	76
	Egg laying to the last minute	76
	Initiation of queen cells	77
	How do the eggs get into the queen cups in the first place?	78
	Special eggs	79
	Who decides that an egg in a queen cup should be allowed to hatch and who does the initial feed of royal jelly?	80
	Which bees in the colony 'know' they are going to swarm and make advanced preparation for S-day?	81
	What changes take place in the colony in the lead up to swarming?	81
	Work to rule	82
	Honey carried by the swarm	83
	Timing of the swarm	83
	The process of swarm initiation	83
	Which bees control this process (initiate the swarm)?	84
	How does the population of the colony split when it swarms?	85
	Natural swarm vigour	87
	The composition of a swarm	88
1.7	**Cast swarming**	89
	An overview of cast swarming	89
	The sequence of cast swarming	90
	The virgins emerge	91
	Cast swarms with multiple virgin queens	92
	Cast swarming in context	92
	Why do colonies cast swarm?	93
	What is adaptive about cast swarming?	93

	Who organises the cast swarm?	94
	New organising bees?	95
	Queen mating	96
1.8	**Selection of the new queen and kin relationships**	97
	The normal colony	97
	What is the evidence for nepotism?	98
	After the prime swarm	99
	The difference between cast swarms	100
	What is the role of queen piping?	101
	Behaviour of cast swarms with multiple queens	102
	Choice of a new queen by the parent colony	104
	Undersized queens	105
	Choosing the queen	107
	Behaviour in the presence of queen cells	108
	Management of cast swarming	110
	Releasing virgin queens	110
	Should the beekeeper interfere?	112
	Conclusions	112

Part 2: Practical swarm control

2.1	**Practical swarm control**	115
	Types of swarm control	116
	Queen clipping	116
	Routine inspection for swarm control	119
	Opening the hive	120
	Weekly hive inspections	122
	Swarm control and honey production	122
	Alternative swarm control strategies	123
	Use of bait hives	124
	Triggers for swarming	124
2.2	**Pre-emptive swarm control**	127
	Accompanying diagrams	128
	2.2.1 Comb management	129

2.2.2	Box management	132
	Creating space for the queen to lay	136
	Other hive configurations	137
	Supering	139
2.2.3	Brood relocation	140
	Downsides	142
2.2.4	Splitting colonies	144
	Timing	145
	Recombining splits	145

2.3 Re-active swarm control (preliminary investigation and diagnostic tree) — 149

Taking control		150
Colony diagnosis		150
Reasons for queen cells being present in the hive		151
Ambiguous situations		152
Swarming or supersedure?		153
Swarm and emergency cells together		154
Some other basic facts you need to know		154
Queen cell development		155
Worker brood development		157
Drone brood development		158
Chronological diagnostic tree and remedial management		159
Quick inspection for queen cells		160
Step 1	There is drone brood in my colony	161
Step 2	There are queen cups in my colony	162
Step 3	There are queen cups with standing-up eggs in them	162
Step 4	There are queen cups with contents (larvae and royal jelly) in my colony and some of the cells are starting to be extended	163
Step 5	There are sealed queen cells in my colony	164
Step 6	The colony has definitely swarmed and is left with a reduced number of bees, brood and numerous queen cells	166

SWARMING BIOLOGY AND CONTROL

	Step 7	My colony has swarmed and there are emerged and sealed queen cells present	167
	Step 8	I think my colony has recently produced a cast (secondary) swarm	169
	Step 9	My colony has no unsealed brood, a limited amount of sealed brood and no sealed queen cells	170
	Step 10	My colony has no brood and no sealed queen cells – help!	170
	Step 11	My colony has no brood apart from that on a 'test' comb that it may have received previously, but NO queen cells have been produced	171
	Step 12	My colony has got a drone-laying queen	172
	Postscript		173
2.4	**Re-active swarm control (artificial swarming and associated management)**		**175**
	Diagram conventions		176
	Using a queen excluder		176
	Determining the stage in the swarming process		176
	2.4.1	The colony has not yet swarmed	177
		Artificial swarming	178
		The first manipulation	179
		Losing the flying bees	182
		Emergency queen cells	182
		The second manipulation	183
		Check for the presence of the queen	183
		Transfer the queen	184
		The emergency queen cells	184
		Artificial swarming using a split board	185
		Looking for the queen	185
		Advantages of Snelgrove II (modified)	188
		Got the timing wrong?	188
		The aftermath of artificial swarming	189
		A two-tier hive	189
		Uniting colonies	191

2.4.2	The colony has already issued the prime swarm but has not cast-swarmed	191
2.4.3	The colony contains both emerged and sealed queen cells and may (or may not) have issued a cast swarm	192
2.4.4	The colony appears to be queen-less (it has no brood) and the beekeeper has no idea what happened and when	192
Late-season swarming		193
	What to do with the queen cells?	194
	A new method of dealing with a late-swarming colony	195
	The explanation	197
Conclusions		197

Appendices

Appendix 1	It is not clear whether the queen cells are for swarming or supersedure	199
Appendix 2	Recommend logistics for Snelgrove II (modified), the second manipulation	200
Appendix 3	Fault finding	202
	Known causes of failure	203
Appendix 4	Failure of the prime swarm	204
Appendix 5	Bait hives – setting up and management	205
	What do scout bees look for when selecting a nest site?	206
	What to put in the bait hive	209
	Signs of impending occupation	211
	What to do when the swarm has arrived?	212
	What if the swarm has been in residence for several days?	212
	Feeding the swarm	213
Conclusions		214

SWARMING BIOLOGY AND CONTROL

FOREWORD

When we beekeepers put a colony of bees in a movable-frame hive, we gain control over where these bees live, and we make it easy to open their home and see how they are doing but, all in all, we gain limited control over the bees. They still determine for themselves where to find the richest foraging opportunities, when to start and stop building their beeswax combs, and how much water to fetch home for cooling their nest or quenching their thirst. A skilled beekeeper can, however, exert considerable control of the swarming by his/her colonies.

These swarm-control methods, and how they are linked to the natural biology of honey bees, are the subjects of this book.

I expect that this book will be welcomed widely by beekeepers, for many of us are keenly dismayed when we witness the emergence of a swarm from one of our hives. This is an exciting display of the bees' behaviour, but often it is also a disappointing sight. It dashes one's hopes for a big honey harvest from the colony that has swarmed because now its forager force is greatly diminished. Moreover, swarming often happens just when one wants the bees to be out gathering nectar, because a honey flow is underway. What the day before was a hive boiling over with bees is now one whose population has shrunk, especially in the honey supers. Foraging is nearly at a standstill, even though the hillsides, fields, riverbanks and woods may be aglow with the blossoms of heather, oilseed rape, Himalayan balsam, or lime trees.

When we watch thousands of swarm bees tumbling from a hive, we are reminded that honey bees are still untamed creatures, ones over which human beings have not totally triumphed. If the swarm has clustered on a low branch or bush, we can then shake the beard-like mass of bees in front of an empty hive almost as casually as if it were a bucket of beans. Soon a few bees will approach the hive's entrance, recognise its quality as a home site, and attract others by raising their abdomens high to release the assembly pheromone, fanning their wings to disperse this chemical 'Come hither' signal. Others bees will receive this signal and will start producing the pheromone, too. Soon, we will watch spellbound as a throng of bees, including the queen, streams into the hive.

Maybe, though, the swarm has settled high in a tree, so capturing it would require risking a broken neck. In this situation, it will be wisest to let the bees find a new home by themselves. We may feel despair; the bees are fine. They are still as much at home in a hollow tree as a manufactured hive, so the step back to living wild is for them a short one.

Soon scout bees are flying forth, discovering possible dwelling places, and sharing the news of their discoveries by performing waggle dances on the surface of the swarm cluster. Eventually, usually within a day or two, the scouts complete their democratic choice of which of a dozen or more possible home sites is the best, whereupon they inform all the other bees to warm their flight muscles, and then the scouts give the signal for everyone to launch into flight. The swarm cluster quickly disintegrates and forms a cloud of swirling bees that begins to fly away. At first it moves slowly, then it gathers speed, and soon it is gone from sight.

As the bees head off to a tree cavity in the woods or a nook in somebody's roof, we can bid them farewell, taking pleasure in the knowledge that these insects have not yielded their nature to us, and are following a pattern of behaviour set millions of years ago.

In Part 1 of this book, Walter Shaw reviews what has been learned about the biology of swarming by honey bees, and what remains mysterious about this special part of their biology. Then, in Part 2, he describes and analyses the panoply of swarm control methods that beekeepers have devised to keep a strong colony from casting a swarm and thereby losing its prolific queen and most of its workforce. Every beekeeper who desires sizeable honey crops from his/her bees needs to understand how a colony decides to swarm, and how to prevent colonies from doing so. This book is, therefore, an important read for most beekeepers.

Thomas D Seeley
Cornell University

Author of Honeybee Democracy *and* The Lives of Bees

SWARMING BIOLOGY AND CONTROL

INTRODUCTION

You could say that this book was born out of an obsession with sex – in honey bees I hasten to add. It is through the process of swarming that a honey bee colony reproduces, so in this context the association of ideas is not inappropriate.

Although parts of it have been written for some time, I have been putting off finishing this book for a number of years. This is because there is always something new to be discovered and each season I find that my understanding of swarming and swarm control has advanced. For me, and doubtless for many others, this is one of the attractions of beekeeping; that there is always more to be learnt and the learning curve never seems to flatten. However, there comes a point where procrastination has to cease and you have to commit your understanding to paper, knowing that it is incomplete (and possibly some of it incorrect), but doing the best you can. That point has now been reached so, for better or worse, here goes.

Beekeeping is a difficult craft to learn because it is so different from anything a beginner has previously encountered. Initially, the novice has no framework on which to hang the information to which he/she is exposed, either verbally or through reading books or articles on the subject. It requires a certain amount of practical experience of inspecting and handling bees before things start to make joined-up sense. I was amazed how, after reading books early in my beekeeping experience, and thinking I understood what they said, re-reading them a few years later revealed things that I had previously not fully understood and it was only then that they started to click into place. So, if you are a complete beginner reading this book, please bear this in mind – it will all make more sense in a few years' time.

My wife, Jenny, and I started beekeeping at the same time and have worked together ever since. If you are a novice, I hope Chapter 1.1 may help you get into the swing of things because it traces our history in relation to swarm control. It starts off in the apiary with us seeing a swarm for the first time and having very little idea what to do about it. A more difficult situation still was when we opened a hive and encountered queen cells for the first time. I will freely confess we panicked, did entirely the wrong thing and destroyed them in the hope this would solve the problem. It didn't and this was the beginning of a long and sometimes frustrating learning process.

This book is divided into two parts. Part 1 is about the biology and ecology of swarming. Personally, I think it is important to understand what the colony is doing and why, and, as far as possible, how it is doing it (the mechanism). This is a fascinating subject in its own right, but also helps the beekeeper make better decisions about the management required to control swarming and also to understand what has gone wrong if things do not work quite as expected. Bear in mind that when things go wrong it is often a more valuable learning experience than when they go right – that is providing you don't lose your patience but take the trouble to work out why things went wrong.

Part 2 is more practical and covers the colony management that can be used to control swarming to the advantage of you as a beekeeper, (hopefully) without compromising the interests of the honey bee colony. At first sight some of the manipulations may sound like hard work and somewhat complicated but when you have done them a few times I am sure you will find this is not the case. Don't be ashamed to take notes (a crib sheet) or even a book into the apiary with you. Also think through the logistics of the intended manipulation before you start and ensure you have all the equipment that may be required to hand. I recognise that not all beekeepers will have the time (or perhaps the ambition) to practice comprehensive swarm control and will have to 'cut the cloth to measure'. Possible strategies in relation to swarm control are discussed in Chapter 2.1.

INTRODUCTION

For the past 20 years we have been running about 50 colonies and this has given us the luxury of being able to experiment, saying, 'what would happen if we did this' and then going ahead and doing it in the knowledge that, if the worst came to the worst, we could afford to lose that colony. We have learnt a lot this way and only rarely paid the price of losing a colony. Several of the techniques described in this book have only been discovered through a willingness to engage in this kind of experimentation which has often been contrary to the generally accepted advice on the matter in hand.

There are a number of new ideas in this book – some of them breaking with the previously published wisdom in relation to swarming – and it will be interesting see how readily they are accepted and survive the test of time. Only time will show!

In summary, it should be emphasised that there is no doubt that swarm control is simultaneously the most important and most difficult aspect of colony management with which beekeepers have to deal – if they chose to do so, of course. It is also the most complex of the many behaviours that have evolved to enable a honey bee colony to successfully complete its life cycle.

SWARMING BIOLOGY AND CONTROL

PART 1

THE BIOLOGY OF SWARMING

SWARMING BIOLOGY AND CONTROL

1.1 OUR EXPERIENCE OF SWARMING AND SWARM CONTROL (A VOYAGE OF DISCOVERY)

Prior to starting beekeeping I had encountered a honey bee swarm just once. A colleague in the research station where I worked was a keen beekeeper and kept a few colonies in the grounds. One of these had just swarmed and, very nervously, dressed in an all-enveloping motorcycling suit, I helped him catch this swarm – well more watched than helped, if truth be known. As is normally the case with swarms, the bees were pretty placid so concerns about my personal safety were unfounded.

My wife (Jenny) and I started beekeeping in 1987 and together have gradually developed our interest and methods over the past 34 years.

The original purpose for keeping bees was to provide better pollination for a newly planted orchard of about 70 trees. We got through our first year of beekeeping with just one colony and, as this made no attempt to swarm, our initial impression was that swarming was not a problem. In the second year, this colony did swarm and, rather inexpertly, we managed to catch the swarm and hive it. I am not sure we even knew that most colonies produce a cast swarm (secondary or after-swarm) a few days later, much less how to prevent this from happening. This first swarm was actually welcome because we now had two colonies which is what we wanted. In the third year, both colonies swarmed and this marked the beginning of a positive deluge of swarming.

Early on, and with some help from a beekeeping friend, we managed to artificially swarm one of our colonies that had set up to swarm using the so-called Pagden method (the method that is still most commonly taught and can be found in most beekeeping books). As luck would have it, this was successful. Over the next few years, regular inspection of colonies during

The Pagden method

The method of artificial swarming that goes under the name of Pagden (James Pagden) is not the one he published in his booklet dated 1868. This dealt with the management of a colony after it had already swarmed and how to prevent the remaining colony from cast swarming.

As his method was described in the context of skep beekeeping (where there is no means of comb inspection), Pagden would have had no way of knowing that a colony had started queen cells, so artificial swarming would have been impossible. This means that he could not have developed the method that bears his name – unless he moved onto movable frame hives at a later date and never published. So when and how his name got attached to the method that is widely used today is a mystery, and will probably remain so.

A swarm in an apple tree

the late spring and summer enabled us to detect those that had set up to swarm and attempt to deal with them.

Again, we used the only thing we knew, the Pagden method, but started to get mixed results. Sometimes it worked and sometimes it didn't, the failure always being that the artificial swarm (with the old queen present) did not settle down to rebuild as the books claimed it would. All too often it would resume the urge to swarm. This could happen within the first few days or, in rare cases, up to three weeks later.

There are a few people who acknowledge the possibility of this outcome (but most don't) and it is usually referred to as 'secondary' swarming. However, because this swarm is headed by the old queen (not a virgin), this should not be confused with cast swarming.

We tried all manner of variations on the method; with and without adding a frame of sealed brood to the artificial swarm and using all drawn combs rather than foundation (or a mixture). None of these variations seemed to make a scrap of difference.

On the plus side, the parent colony (the one containing the brood and the queen cells but not the queen) on a new stand never gave any problem. Perhaps fortunately, we did not realise until later that some books said that you had to reduce the number of queen cells in the parent colony down to one to prevent that from swarming. We never did this and the colony never swarmed, thus confirming it is an unnecessary precaution and the explanation for this will be discussed in Chapter 1.6.

As far as the artificial swarm was concerned, there did not seem to be any rhyme or reason about the outcome and, based on what was said in beekeeping books, we came to believe that either we must be doing something wrong (but what?) or that our bees were excessively prone to swarm and did not obey the normal rules.

The truth begins to dawn

Other beekeepers we talked to claimed to have little problem with swarming and this made us feel even more inadequate. It was not until we started to visit other people's apiaries that the truth began to dawn. One apiary we visited contained 11 colonies none of which, according to the beekeeper, had swarmed. Inspection on the day showed that nine of them either had swarmed or were about to do so at any moment and the two that hadn't were too weak to do anything so reckless. Other similar visits made us realise that many beekeepers had little knowledge about the state of their colonies and that many rarely looked at anything below the queen excluder.

Louis Snelgrove

About this time, a non-beekeeping friend purchased for us (for the princely price of 10p in a bring-and-buy sale) a rather battered copy of Louis Snelgrove's book, *Swarming: its Control and Prevention* first published in 1934. This was the sixth edition of 1942 which contains some changes, probably introduced in 1936. A later fifteenth edition, re-published by Bee Books New and Old in 1998, contains some further minor changes.

Most of the standard beekeeping books contain some information about swarming and its control but this varies greatly both in the amount of detail and accuracy.

On reading some books I often suspect that the author is just repeating what has been written by someone else and lacks first-hand, practical experience – alternatively they are being economical with the truth. On the other hand, Snelgrove obviously had many years of experience dealing with swarming and had approached the problem using careful observation and in an analytical frame of mind. He had tried many different methods and his book contains a lot of useful and thought-provoking information.

Snelgrove's book opened new doors on the problem of swarm control. Being an enthusiastic woodworker, I made several split boards to his design and we started to experiment with them.

A Snelgrove board

With his book in hand (literally), we worked our way through what he called Method I, which is for use with colonies that have not yet set up to swarm but had reached a state of development in which they had the potential to do so in the near future.

We quickly realised that to follow the fairly complex instructions concerning the use of the paired doors on his board was unnecessary or, in some cases, undesirable.

> Split boards
>
> A horizontal split board is simply an intermediate floor that is used to divide a hive vertically, creating two separate colonies, one on top of the other. Such boards have been around since the early days of the movable frame, modular hive and they are usually named after their inventor (eg, the Wedmore board or Horsley board). Common features are that they have at least one entrance on the top side of the board.
>
> All the designs of split board of which I am aware have a central hole that is covered with a piece mesh which can be removed when required. The mesh allows warmth to come up from the colony below and also a limited amount of communication between the two colonies (bees can antennate each other through the mesh).
>
> The Snelgrove version is an extremely elaborate board, having no less than six or eight entrances in vertical pairs communicating with both sides of the board (see page 7). Strategic opening and closing of these doors can be used to adjust the number of bees in each colony by directing returning foragers into either the upper or lower box. However, for most methods of swarm control, a much simpler split board with a single entrance on its upper side is all that is required (see page 9).

A split board made from a modified cover board with a single entrance at the front

Snelgrove based his methods on the Gerstung theory of swarming, first published in 1890, and it is from this that most of the complexity arises.

The Gerstung theory

The Gerstung theory claims that swarming is due to an excess of brood food produced by the nurse bees compared to the amount of brood that requires feeding. This condition occurs when colonies are approaching their peak size, when there is a large population of nurse-age bees (with new ones emerging all the time) and the queen is beginning to lay less. This nicely coincides with the conditions that prevail when most colonies actually do set up to swarm – so the theory seems to make sense. Also, what is the first sign of a colony's intention to swarm that the beekeeper sees? It is queen cups with larvae floating in generous pools of royal jelly. The circumstantial evidence was rather compelling but was not supported by any hard evidence.

The Gerstung theory held sway for about 60 years, although it did have some detractors and it is clear that Snelgrove was actually aware of this when he wrote his book.

In the 1950s, Dr John Simpson in the Bee Department at Rothamsted Experimental Station (now Rothamsted Research) carried out some simple but elegant experiments to investigate the Gerstung theory. As soon as the queen had laid in them, he removed as many frames with eggs from the colony as possible and replaced them with frames of sealed brood from another colony. In this way he created a colony which contained a huge number of nurse bees but virtually no brood for them to feed. The colonies treated this way did not swarm, thus providing incontrovertible evidence that the Gerstung theory was incorrect. This means that many of the details in Snelgrove's instructions that aim to control the distribution of the nurse bees between the two colonies are totally unnecessary.

Snelgrove's Method I

Method I can now be seen as simply a method of splitting a colony vertically as opposed to horizontally, thus avoiding the need to utilise a new hive stand.

Splitting colonies to prevent swarming has a long history going back into the nineteenth century, so there was nothing new there – this was just a convenient and flexible way of doing it. The paired doors on the Snelgrove board enable the beekeeper to divert flying bees from the top part of the split into the bottom colony, thus strengthening the part that contains the queen and is destined to become the main honey producer.

We found that the best strategy for the use of the paired doors was to wait 5–7 days, until all the bees with previous flying experience had returned to the bottom part of the split, and then examine the top part. On the basis of this inspection we would decide whether or not this part of the colony could afford to lose bees by changing the doors that were open. This is discussed in more detail in Part 2.

Just as happened to Snelgrove, we misjudged some colonies and they started swarm cells before we could split them. In this situation, initially we followed his modification of Method I (for colonies that had developed queen cells) which is simply a vertical version of the Pagden method. Not surprisingly, this proved to have the same unreliability as the horizontal version and had the disadvantage that it made it more difficult to check what the artificial swarm was doing.

Even if it is possible to detect the artificial swarm's intention to swarm again (before it actual does so), it is difficult to know what to do about it. The only measure possible is to remove the old queen (put her in nucleus box), remove any sealed queen cells, wait until there is no possibility of the bees making any more and then reduce the number to one. So in a sense we were back to square one of being unable to artificially swarm colonies and be assured of a reliable outcome.

Snelgrove's Method II

Reading further into Snelgrove's book we found what is called his Method II. This differs from Method I (and Pagden) in the fact that when the colony is split the queen remains with the brood and queen cells (the parent colony) and is only later transferred to the artificial swarm consisting of the flying bees and a full set of combs including one frame containing sealed brood.

To most beekeepers, leaving the queen with the queen cells is counter-intuitive and sounds like a recipe for disaster. Their reasoning is that you have changed nothing and a colony with both a queen and queen cells present is bound to swarm. But you only have to try it to discover this is not the case. It is because the flying bees have relocated to the artificial swarm that this manipulation causes the bees in the parent colony to break down the queen cells and make no attempt to swarm.

Destruction of the queen cells does not occur immediately but only when the pupae have reached about day 12–13 in their

Snelgrove's Method II

Snelgrove's discovery of Method II is a perfect example of serendipity (a happy accident).

In one of his apiaries, several colonies had caught him by surprise and started queen cells before he had split them according to his Method I. This forced him into artificially swarming them using his modification of Method I (effectively a vertical Pagden).

When later he inspected these colonies, he found that in two of them the queen in the artificial swarm had managed to squeeze her way through the queen excluder into the supers.

Because it was his practice to leave the mesh panel out of the board for the first 48 hours (aimed at allowing a redistribution of the nurse bees), the queens were able to find their way into the box containing the brood and queen cells at the top of the hive.

When, a few days later, he inspected this box, he was amazed to find that the queen cells had been torn down and the queen had resumed laying. He immediately realised the potential of this behaviour in relation to artificial swarming. So to check this had not been a one-off event, he artificially swarmed several other colonies, but this time he deliberately left the queen with the brood and queen cells and this confirmed his initial observations.

It is interesting to note that, in his instructions for what he now called his Method II, he said that the artificial swarm should be left with one frame of sealed brood and on no account anything from which the bees could make queen cells.

Hold on to this thought and read on.

development. It is only at this point that the bees become fully aware of their presence and (using human logic) realise that they already have a perfectly good queen so why do they want another?

If the queen cells are in a fairly early stage of development when artificial swarming is done, the delay in tearing down the cells unnerves some beekeepers. It is just a matter of keeping faith because, in our experience, it will happen at the appropriate time without fail. Also, as soon as the flying bees have departed to join the artificial swarm, the queen resumes laying normally.

In a second manipulation 9–10 days later, the queen is transferred to the artificial swarm which will happily accept her and usually makes no attempt to swarm. Meanwhile, back in the parent colony, when the queen has been removed, the bees will react to her loss by immediately starting emergency queen cells based on the queen's recent laying. In due course the colony will automatically re-queen itself with no need for any further intervention by the beekeeper.

Having accumulated all the flying bees, the artificial swarm remains the main honey producer and the paired doors on the Snelgrove board can (judiciously) be used to supplement the foraging force. The biological explanation for how and why Method II works in the way it does will be given later in context, but a clue has already been given – it is control of the flying bees. Method II proved to be a great improvement over Pagden but frustratingly still did not give 100% reliability.

Bees in the artificial swarm finding the queen

At this point we encountered another problem. In his book, Snelgrove warned that when he used Method II, the bees in the artificial swarm occasionally found where the queen had gone and moved up to join her in the top box. In practice we found this happened a bit more than occasionally and, if not noticed,

SWARMING BIOLOGY AND CONTROL

it would not be long before a swarm issued from the parent colony on top of the split board.

If the beekeeper is there when it starts and spots it, it is pretty obvious what is happening, with a stream of bees walking up the side of the hive to enter the entrance on top of the split board in order to reunite with the queen. In an out-apiary, where colonies cannot be inspected on a daily basis, this shortcoming is unacceptable.

However, providing it has been seen in time, it is easy enough to recover the situation. To prevent swarming, the parent colony on its board (acting as a substitute floor) should be moved to a new position in the apiary. The flying bees, which are the source of the problem, will return to the artificial swarm in its original position and will be unable to re-find where the queen has gone.

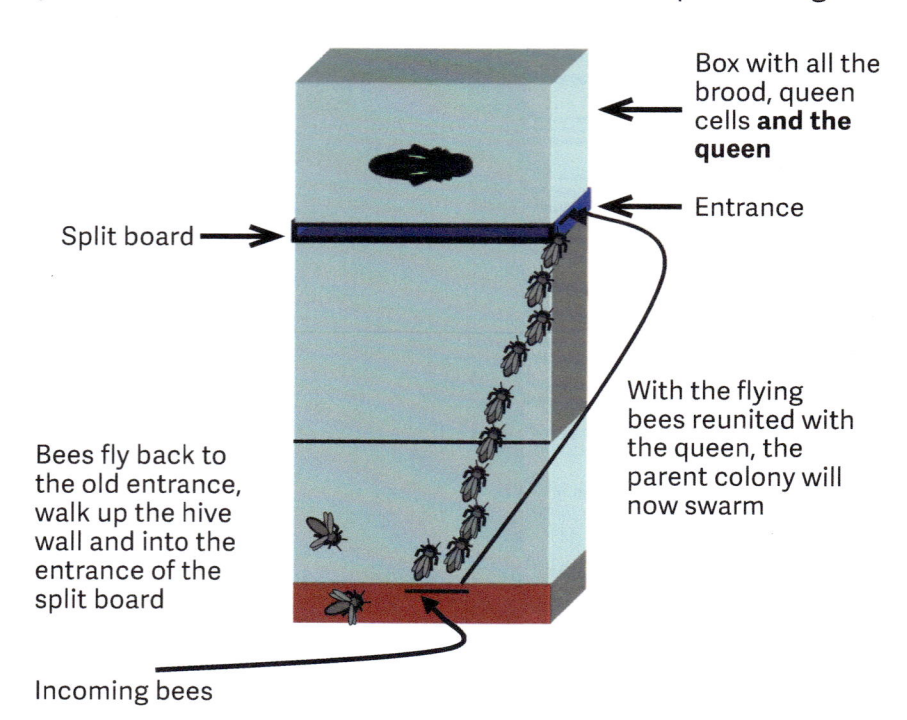

Box with all the brood, queen cells **and the queen**

Entrance

Split board

With the flying bees reunited with the queen, the parent colony will now swarm

Bees fly back to the old entrance, walk up the hive wall and into the entrance of the split board

Incoming bees

Worker bees from the artificial swarm moving up to join the queen in the top box

This is fine if you have a spare hive stand and want a new colony in that position but, if the plan is to unite the two colonies some time later in the season, it creates logistical problems. It should also be noted that further equipment is involved because the artificial swarm now requires a replacement cover board and roof.

Continuing failure

Even when everything ran smoothly, and the bees did not find where the queen had gone, there was still a failure rate with Method II. After the queen had been returned to the artificial swarm we found it could still swarm, but less frequently than when using the Pagden method.

Although not acknowledged in his book, I suspect that Snelgrove was aware of this because he went on to devise a Method IV. We have never tried this method – which I doubt solves the fundamental problem of artificial swarming anyway and also has some fairly obvious downsides. If anybody is interested and wants to give it a go, it can be found in his book.

A Eureka moment

As a result of one of those 'what if?' moments that I mentioned in the Introduction, a couple of seasons later we found a solution to the problem of the bees finding where the queen had gone.

Instead of the artificial swarm having a frame of sealed brood, we gave it a frame (or two frames) of brood containing eggs and newly hatched larvae. (NB. This is specifically contrary to Snelgrove's instructions.) With no queen in its midst (because she is in the parent colony), this enables the artificial swarm to make provision for a replacement queen by starting emergency queen cells – which it does with great alacrity. The idea is that this offers the bees an 'escape route' from what they must recognise as a hopeless situation (no queen and no means of making one).

The original thought behind the move was that it would distract the artificial swarm from trying to find their lost queen – but this may be 'human logic' which is not necessarily the same as that of the bees.

This was a Eureka moment! It was successful and did what it was intended to do. Not only did it prevent the bees in the artificial swarm from wanting to find out where their queen had gone, but we soon realised it was having a more profound effect on the artificial swarm, 'effectively' eliminating the urge to swarm. Our immediate reaction was to call this method Snelgrove II (modified) – which is what it was.

From the subsequent behaviour of the artificial swarm, we reasoned that this is because it has been switched from a swarming programme onto an emergency re-queening programme. Of course, using this modification, the frame (or frames) carrying emergency queen cells must be removed at the same time that the queen is transferred to the artificial swarm on day 9–10, otherwise it is not clear what would happen. What to do with the removed frames bearing queen cells is discussed in Chapter 2.4.

Method II (modified) – total success

With Method II (modified) we found we had discovered a very simple method of artificial swarming which had a 100% success rate and we have used this ever since (for about 15 years with some 200 successful replications under our belt).

Because it was the simplest thing to do and out of deference to the great man, we used his name from the outset. Unfortunately this has caused some confusion with people not understanding the importance of the word 'modified'. In hindsight, the name was probably a mistake and some people (rather embarrassingly) call it the Wally method. This method will be described in detail in Chapter 2.4 (using colour-coded diagrams) along with further explanation as to why it is thought to work so much better.

'Effective' elimination

An observant reader, or one with a suspicious mind, may have noted that the word 'effectively' is enclosed in quotation marks.

In normal use, the method (as described above) completely prevents any attempt by the artificial swarm to resume swarming. However, in the case where a colony is found to have started queen cells in late June or early July, just before the main nectar flow in most areas, the last thing the beekeeper wants to do is split the colony and compromise the foraging force.

So what can be done about it?

There is no alternative to artificial swarming, but on day 9–10, when transferring the queen to the artificial swarm, would it not be possible to reunite the whole colony under one roof? At least this would limit the time the colony was in two separate units. (NB. To attempt this earlier in the season would not be a sensible idea because it would simply recreate the conditions that had triggered the colony to swarm in the first place).

Again, in the spirit of experimentation, we tried reuniting colonies at this time but, a bit like the Pagden method, it proved to be unreliable – sometimes it worked and sometimes it didn't. Clearly this must be because there is still a residual urge to swarm present in the artificial swarm that, given sufficient encouragement, can reassert itself. Hence the use of the word 'effectively'.

Fortunately, this finding did not prove to be a dead end, because we have since found a reliable way in which the colony can be reunited under these circumstances (details are given in Chapter 2.4).

SWARMING BIOLOGY AND CONTROL

> ### The Snelgrove board
>
> At this point it is worth emphasising that none of the methods devised by Snelgrove (including Method II) actually require the use of his board (or any other type of split board). They can all be done horizontally, using a new hive stand (see Chapter 2.4).
>
> When Method II is done this way, because of the spatial separation, it eliminates the problem of the artificial swarm finding where the queen has gone. However, use of a split board does afford an economy of equipment; a second floor, roof and cover board are not required. It also provides logistical advantages if the artificial swarm and parent colony are to be united at some time later in the season.
>
> You may wish to do this for all sorts of reasons, but the bottom line is that, through repeated use of artificial swarming, you can't go on indefinitely doubling your number of colonies – this is the route to madness or divorce!
>
> By a stroke of good fortune, each time we used Method II it involved the use of a split board and this led to an important discovery which changed our understanding of the swarming process – read on!

Swarm prevention and control

As already noted, this book is divided into two parts; the first part covers what is known about the biology of swarming and the second part deals with the practical methods of both pre-emptive and re-active swarm control.

Many beekeeping books (including that by Snelgrove himself) use the words (swarm) 'prevention' and 'control' more or less indiscriminately and this can prove to be very confusing.

Throughout this book the following terminology is used to clearly differentiate between what the beekeeper can do to try and prevent a colony from starting queen cells (**pre-emptive control**) and what can be done to prevent a colony that has started queen cells from actually swarming (**re-active control**).

The appearance of queen cells in a colony is a biological tipping point (threshold) at which one form of management is no longer effective and has to be replaced by something entirely different.

This format of attempting to understand the causes and biology of swarming, followed by the practical methods for its control, has been attempted before in a little known book with the title, *Swarm Control Survey* (but a more appropriate title would have been *A Review of Swarm Control Methods*). The author of this book (published in 1946) was a Hampshire beekeeper, ER Dent. The book has a forward by LE Snelgrove and is a very credible attempt to cover the subject. However, because so much less was known about bee biology at that time, some of the theories that it advances are clearly wrong.

Summary

One of the many sayings you hear from beekeepers is, 'the bees don't seem to have read the same book that I have'. However, this overlooks the fact that it was not the bees themselves that wrote the book. It was us beekeepers who wrote about them and this is where the gap in mutual understanding arises.

A title that I have used for a talk in the past is, 'The bees know what they are doing but does the beekeeper?' The wording is intentionally ambiguous, the obvious interpretation being that the beekeeper does not know what he or she is doing. This implied insult certainly grabs people's attention – but I have yet to be lynched. The correct interpretation is that the beekeeper does not understand what the bees are doing and therefore might not be acting in their best interest – or in some cases against it.

Through natural selection, the honey bee colony has evolved a series of complex and highly adaptive behaviours to deal with all eventualities that it is likely to encounter throughout a life in the wild. A colony comes equipped with these behavioural programmes (or instincts) which seem to be 'hard wired' into its neural network. Within these programmes, the bees are able to use their collective intelligence to ensure they (a) choose the right programme and (b) implement it in a way that gives them the best chance of success. However, through our management, we beekeepers can sometimes accidentally (or even arrogantly) create situations for which the bees do not have a ready-made programme. Alternatively, we may fail to give the bees enough information to help them determine their best response.

The point I am trying to make is that we humans are capable of understanding bees (it may not always be easy) but they are certainly incapable of understanding us. Yes, a colony of bees has a highly sophisticated form of intelligence but it is not capable of innovation (solving novel problems). So, from a practical point of view (and perhaps also an ethical one), it is our responsibility to understand the biology of the honey bee colony as much as we can and to use this as the framework within which we interact with our bees.

Going back to the beginning of this chapter, I stated that our original motive for keeping bees was to provide pollination for our orchard. As we gradually learnt about the craft of beekeeping we came to realise what fascinating and complex organisms bee colonies are. Being retired research ecologists, we had the background and interest to further explore the fast developing field of honey bee biology and try to work out how this knowledge could best be applied to practical beekeeping.

Because of the importance of swarming to both the bees and

the beekeeper, this part of colony behaviour has received a lot of attention. As you can see, our motivation in beekeeping has changed radically over the years but we have not lost sight of the main aim of beekeeping, which is to produce honey. We do produce and sell honey (quite a lot in a good season) and this is a nice little earner which more than defrays the cost of our beekeeping. We also raise new queens and produce starter colonies of locally adapted bees for new beekeepers and have developed methods that have a high level of success. However, this is a service rather than moneymaking enterprise.

Nuclei in a mating apiary

SWARMING BIOLOGY AND CONTROL

1.2 THE REPRODUCTIVE STRATEGY OF THE HONEY BEE

Much of this chapter is involved with the field of evolutionary biology; how and why evolution has shaped the reproductive strategy of the honey bee colony the way it has. Reproduction is the single most important function of the colony, to which all other functions are subservient.

It should be clearly understood that the reproductive strategy of the honey bee is very different from that of most animals with which we are familiar, eg, farm stock and pets. Many attempts to breed bees ignore these fundamental differences. In my view, this is seriously misguided and, because of its reproductive strategy, the honey bee is not a suitable subject for 'domestication' (moulding it to our human requirements) and should be allowed to remain as an essentially wild animal.

The order Hymenoptera (meaning membranous wings) is an extremely large group of insects comprising 91 families and over 115,000 known species, which include ants, bees, wasps, ichneumons, chalcids, sawflies and other lesser known types.

One important feature they have in common is the haploid/diploid method of sex determination. Males are haploid, being developed from an unfertilised egg (by parthenogenesis), and therefore have only one set of chromosomes. Females develop from a fertilised egg and are therefore diploid (with two sets of chromosomes). This enables them to adjust the sex ratios, eg, not having to support males during the time when their services are not required, and is one of the factors that has enabled a social life-style to evolve in some families in this order.

Sex alleles

Although sex determination in the Hymenoptera does not involve X and Y chromosomes (as in mammals), there are genes called sex alleles and these come in a number of variants.

Haploid/diploid drones

If a bee inherits just one type of sex allele it is male. A drone is haploid so can only have one type and must therefore be male. A problem can arise with a fertilised egg, which gets one allele from the queen mother and one from the drone father.

If these are the same (matching sex alleles) then this offspring automatically becomes male – known as a diploid drone. On hatching, such larvae are immediately identified by the workers as 'undesirables' and destroyed. This ensures there are never any diploid drones in a colony. However, they are technically viable and can be raised artificially in an incubator. Diploid drones raised in this way are capable of producing a small amount of sperm but it is also diploid and probably non-viable.

If a queen has mated with a drone that carries a sex allele that matches one of hers, 50% of the eggs that result from that pairing will be non-viable. This can be seen as holes (empty cells) in the brood pattern but the beekeeper should not jump to the conclusion that these are the result of diploid drones because there are other possible causes.

The most common explanation for holes is that the queen simply missed laying an egg in a cell. Some queens seem to be less methodical in their laying than others but whether or not this should be considered a fault is open to question.

To properly assess the laying pattern of a queen it is necessary for her to have access to a comb with a substantial area of cells which have been prepared for her to lay in. When only one or two out of a total of 10–20 drones with whom a queen has mated has a sex allele that matches one of hers, there will be gaps in the brood pattern but not enough to be of great significance.

If allele matching is more frequent, it creates a condition called pepper-pot brood (widely scattered empty cells) and, in this case, the loss of potential workers can prevent the colony achieving its full potential. However, this is not the only cause of pepper-pot brood and it could be a symptom of brood disease which needs to be checked.

From this it can clearly be seen that the last thing that a virgin queen wants to do is to mate with a drone produced by her own mother but, of course, it can occur by chance. Late-season queen supersedures (where the colony replaces its queen without swarming) run an increased a risk of this happening.

Drone production

Despite the fact that I don't think a colony will ever set up to swarm before it has produced some drones, the old idea that a colony starts to produce drones so that it can swarm should be dismissed as one of those beekeeping myths.

A colony produces drones because this is an alternative way by which it can pass on its genes – of course, the more obvious way is through the queen producing daughter queens.

SWARMING BIOLOGY AND CONTROL

Drones and drone brood

Producing drones and feeding them requires a lot fewer resources than swarming, so it becomes a viable option earlier in the season. Even though a colony has its own drones early in the season and before other colonies in the area, a premature attempt to swarm is unlikely to be satisfactory because the new queen will be unable to access the diversity of drones with which she needs to mate to become a 'good' queen – one that produces workers with a wide range of genetic variation.

This is often the fate of early-season attempts at supersedure but this may be the colony's 'last desperate shot' at survival so there is little that can be done about it. At least a poorly mated queen is usually good enough to produce a daughter so there is a chance that the genetic line will be preserved by this means.

Eusocial insects

Although the vast majority of species are solitary, the Hymenoptera includes most of the known social insects, including ants, bees and wasps. Only the ants, honey bees and some of the stingless bees are fully eusocial, ie, maintain a co-operative colony that persists throughout the year and from year to year.

Queen longevity

Honey bee queens have evolved to live for up to four to five years to give them the maximum chance of passing on their genes in that time, but some ant species have taken this evolutionary pathway a lot further and have queens that can live for up to 30 years and it is a remarkable feat that they are able to store viable sperm for this amount of time.

The queen in a honey bee colony is usually replaced during swarming but if the colony recognises the need to replace her at a time when swarming would not be possible (or ill advised), they can do this through a process called supersedure. This involves the production of a limited number of queen cells (usually no more than three) with no intention of swarming. After a swarm has established itself at a new nest site it will often decide to replace an aging queen using the same method.

Contrasting reproductive strategies

Although honey bees and ants both evolved from a common wasp-like ancestor, they have totally different reproductive strategies which can be summarised as **quality** as opposed to **quantity**.

Honey bee colonies invest a considerable proportion of their resources into producing just a few offspring each year; the prime swarm and a maximum of about three cast swarms. A honey bee queen is so specialised (just an egg-laying machine) that she requires the support of a large number of workers to be able to fulfil her role, so the provision of a swarm (her support team) is an essential part of reproduction. This strategy is called K-selection, in which each offspring receives a large input of resources and has a relatively high probability of success.

By contrast, when an ant colony is mature it produces a large number of reproductive individuals – winged males and females. Under suitable weather conditions these issue from the nest, take to the wing and mating occurs usually on the same day. By some little understood mechanism, most colonies of the same species in a locality do this on the same day (reproduction is synchronised) and this ensures cross-mating. A large number of mated queens are produced, disperse and individually (without any help from any worker ants) have to establish a new nest for themselves. This is called R-selection in which the inputs are small; each queen has a low probability of success but there is large number of them (strength in numbers).

Plant extremes

The same but even more extreme reproductive strategies also exist in the plant world, with the coco de mer that produces the largest seed in the world of up to 40 kg and an epiphytic orchid that produces seeds that weighs about 10 µg (10 millionth of a gram).

The Feminine Monarchie

In 1609, the Revd Charles Butler published his famous book, *The Feminine Monarchie*, in which the role of the queen was for the first time fully recognised as being the mother of the colony. Prior to that the different-looking bee which all colonies were known to contain was usually referred to as the king.

Butler lived in a male-dominated human society so what else would you call this obviously important individual? But did the Revd Butler get it right? Strictly speaking he didn't, because in biological (genetic) terms, the honey bee queen is

hermaphrodite – both male and female. This is because the drones that she produces are developed from unfertilised eggs and only carry a half-selection of her genes, with no contribution from the drones with which she mated. This means they are half-clones of their mother and their role is to replicate her genes in the form of several million sperm, each containing identical genes.

The drone also acts as a delivery system on behalf of the queen and thus is functioning as her male genitalia. What is more, drones are flying genitalia (the mind boggles!) thus enabling the queen to disperse her genes widely and be in more than one place at the same time – a good trick if you can manage it.

Following this to its logical conclusion, we should perhaps re-name the queen as the 'quing' (not a serious suggestion!). The fact that the queen is hermaphrodite has no practical significance for beekeeping but it does mean that we should not think of the drones in a colony as brothers to the workers. In fact, their genetic status means that they are more akin to uncles.

Maximising genetic diversity

The overall reproductive strategy of a honey bee colony is out-breeding in order to maximise the genetic diversity of its workers. To achieve this the queen practices polyandry (mating with many drones).

Honey bee virgin queens have been shown to fly up to three miles from their nest site in order to ensure they access as much genetic diversity as possible and also reduce the chance of mating with drones produced by their own mother. Polyandry also occurs in other species but the honey bee is the most extreme example known to science and, as a result, it is now often referred to as an example of hyperpolyandry.

The frequency of mating and the distance which queens fly to attend drone assembly areas (DCAs where drones from

Drones attracted to a tethered queen in a drone congregation area

the surrounding area congregate awaiting the arrival of virgin queens) has obviously been determined by evolution to ensure optimum genetic diversity. This behaviour clearly involves risks but must be adaptive (confers considerable advantages) for it to have evolved this way.

For the temperate races of honey bee, the target is for the queen to mate with between 10 and 20 drones, although recent studies using improved methods of genetic discrimination suggest that this figure is higher (15–25 and sometimes more). Some tropical races of honey bee (eg, *Apis mellifera scutellata*) have been shown to have a higher average mating frequency. The giant honey bee (*Apis dorsata*) is thought to hold the record for mating frequency with values approaching 100.

When thinking about mating frequency and the distance that queens fly to mate, we should remember that this has evolved under natural conditions and not in beekeepers' apiaries, where some level of genetic diversity is available locally. Wild colony density has been shown to have an average of about one per square kilometre.

Potential loss of genetic diversity

If the colonies in an apiary are mostly unrelated they will provide a rich source of genetic diversity at short range. However, if through more controlled methods of queen rearing, colonies are closely related, local diversity may be seriously lacking. Where beekeepers use isolated mating apiaries, in which the only source of drones is carefully controlled (by them), this can result in a lack of genetic diversity thus defeating the aim of polyandry.

It should be clearly understood that this sort of practice undermines the evolved reproductive strategy of the honey bee and, if used repeatedly, can result in a serious loss of genetic diversity. The use of instrumental insemination can reduce genetic diversity even more drastically – which regrettably usually seems to be the intention!

Gene cross-overs

Finally, the honey bee has another trick up its sleeve to maximise genetic diversity. When a diploid organism produces gametes (eggs or sperm) the number of chromosomes needs to be halved. This is achieved during the process known as meiosis in which the double-stranded chromosomes unravel and are pulled apart. This does not occur cleanly and sections of the paired chromosomes are exchanged taking the genes they carry with them – these are called cross-overs. This results in enhanced gene mixing making it even more unlikely that two gametes have exactly the same genetic make-up.

The rate of cross-over that occurs when the honey bee produces eggs is 20 times what it is in humans and this is yet another source of genetic diversity. Although not adding any new genes to the colony's pool, cross-overs serve to increase the genetic diversity of the workers. Also, being in a different mix may change the expression of some genes, ie, many genes do not work the same in isolation as they do when they are accompanied by other genes (polygenetic traits).

Because a drone is already haploid, no meiosis occurs in the production of sperm (ie, there are no cross-overs) so all those that he produces have an identical genetic make-up. The function of a drone is simply to replicate the genes contained in the egg from which he was produced, so what matters here is the cross-overs that occur when the queen produces her eggs and this is what creates greater diversity amongst the many drones that she produces.

Because the last thing daughter queens want to do is mate with drones produced by their mother, they do not get any direct benefit from this source of genetic diversity. However, queens produced by other colonies in the area do benefit by encountering these drones (produced by a different queen) when they attend DCAs. The driving force behind the increase in cross-overs may well be an example of what is called group selection (in this case at a species level) – a subject that still creates heated arguments amongst evolutionary biologists.

As already noted, queens flying some distance from their nest site to mate with multiple drones incur an appreciable risk. The fact they take this risk underlines the importance honey bees attach to harvesting maximum genetic diversity. So beekeepers beware; avoid practices that could interfere with the honey bee's evolved reproductive strategy.

The honey bee is a highly successful species that has been around for a long time and it is foolish (even arrogant) of us to think we can better a behaviour that has developed through natural selection. This argument is beautifully summed up in an article by internationally recognised honey bee pathologist

Leslie Bailey, published in *Bee World* (1999) with the title of 'The quest for the super-bee'. This is his summary:

'Highly intensive selection of the honey bee for any quality may decrease its resistance to its wide variety of enzootic pathogens by decreasing its genetic variability. Maintenance of naturally adapted regional strains by traditional means and management that least inhibit their essentially independent lifestyle may be more rewarding.'

You also need to be aware that this uncompromising view was written before the all-embracing benefits (not just disease resistance) that genetic diversity confers on a colony began to be recognised.

Evolution of queen longevity

Queen honey bees must have started their evolutionary development as slightly modified workers which retained the ability to mate and lay eggs. Obviously the longer a queen lived the greater was the chance of her being able to pass her genes on to the next generation, so increased longevity would have automatically arisen through natural selection. This has resulted in honey bee queens surviving for up to four to five years. Even more remarkably, they have developed a method of storing sperm and maintaining its viability for the duration of their lives and also the ability to dispense it so sparingly when fertilising eggs that what is a relatively small volume of semen lasts for the duration.

Studies show that a queen can store up to six million sperm and lay up to 1.5 million eggs during her lifetime, which suggests that fertilisation only requires an average of four sperm/egg. The same volume of semen is released for each egg and, with an older queen who is starting to run out of sperm, this may contain as little as two sperm. This is very different from fertilisation in mammals which requires a large number of sperm in order to be successful.

SWARMING BIOLOGY AND CONTROL

> ### Queen longevity
>
> It can be seen from the above that longevity of queens is a characteristic that is under constant selective pressure – because the longer a queen lives the more likely she is to pass on her genes.
>
> If during large-scale queen rearing or breeding programmes, the selection of 'breeder queens' favours young queens (perhaps the 'best' that were produced in the previous year), this could serve to bypass the natural selection process. Especially if repeated year-on-year, this could result in a gradual decline in longevity.
>
> In the USA, where there is serious concern about queen longevity, many replacement queens are supplied by high-volume producers (usually in the southern States where the climate is favourable). It is interesting to speculate whether this might not be at least part of the problem. In our own small-scale queen-raising programme we do not usually breed from a queen until she is in her third year because we like to have one full season over which to assess the characteristics of our breeder queens.

Queen substance

The honey bee queen has also developed other 'skills', the most important of which is to produce a mix of pheromones (queen substance) that suppresses sexual development in her daughters (the workers). Additional pheromones produced by the brood reinforce those of the queen.

The queen's ability to deny her worker daughters the ability to reproduce, thereby passing-on their genes, is regarded by some authorities as a form of enslavement. However, it is absolutely essential for the harmonious conduct of colony activity to avoid competition for the right to reproduce.

(A similar strategy can be found in some higher animals, eg, wolf packs, where a dominant pair [male and female] are the only ones that are usually permitted to breed. This reduces competition between individuals and enhances co-operation within the pack. But this is only possible at the expense of a good deal of violent enforcement which honey bee colonies seem to be able to avoid.)

For the honey bee colony, the suppression of the ability to reproduce in all but the queen means that the genetic future of all the many thousands of workers is equally vested in their mother (the queen) and they all work solely (and selflessly) in her interest – even to the point of sacrificing their lives in her defence.

Laying workers

All colonies have a small number (about 1%) of workers with functional ovaries which can lay eggs but, because they are unable to mate, these will result only in drones. Even though these drones have only a very slim chance of mating with a virgin queen, thereby passing on their worker-mother's genes, this puts the laying workers in competition with other workers in the colony which would wish to avail themselves of the same advantage.

Worker police

This potential cause of dissent is overcome by the colony's specialised 'worker police' which patrol the brood nest looking for worker-laid eggs which are immediately destroyed. They can probably identify these eggs because they have not been chemically marked in the way the queen does during laying. This clearly demonstrates the lengths to which a colony is prepared to go in order to maintain social harmony. When later we come to consider cast swarms (Chapter 1.8) there will be other examples of how this is achieved.

SWARMING BIOLOGY AND CONTROL

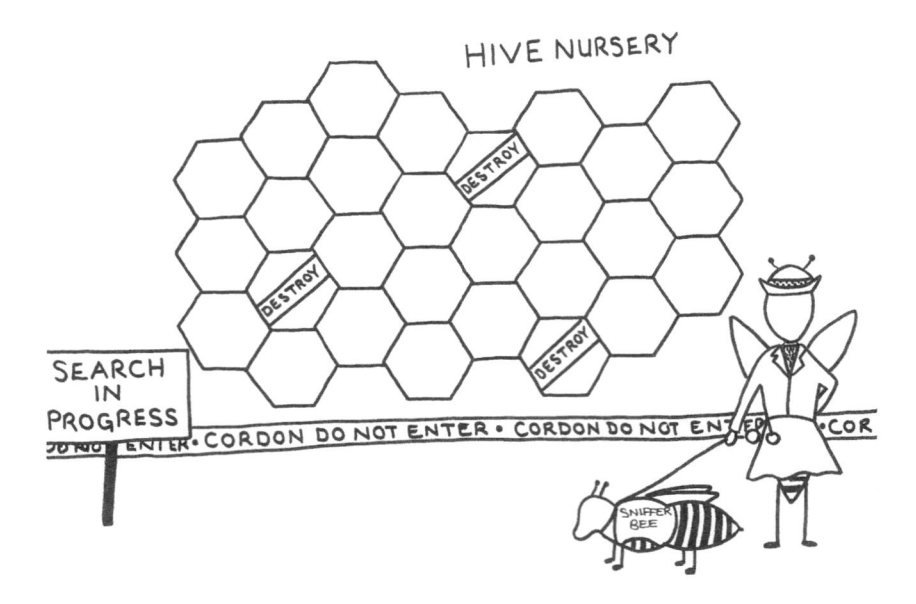

In a colony that has become queen-less, in the absence of a supply of queen substance and gradual diminution of the brood pheromone, it takes about three weeks for the ovaries of some workers to develop sufficiently to produce eggs. They can also produce small amounts of queen substance which may fool the colony into thinking it has a queen and this can make it more difficult for the beekeeper to get a replacement queen accepted.

The presence of laying workers can be detected through their poor laying pattern and the fact that there are often multiple eggs laid in cells. Most books claim that workers are unable to lay eggs at the bottom of the cell (they will be on the side wall) but, in my experience, this does not provide reliable evidence as to who is doing the laying.

Queen mating

Sometime during the first two weeks of her life the queen takes one or more mating flights. How much the queen herself calls the shots is an open question because the workers seem to control much of the process; they hassle her out of the hive under suitable weather conditions, they probably determine how many flights she makes, and the latest studies indicate that she is accompanied by an small escort of workers which guide her to and from the mating venue (a DCA).

> **The Cape honey bee**
>
> In the honey bee family there is just one exception to the suppression of reproduction by workers and that is found in the Cape bee (*Apis mellifera capensis*) – native to the Cape area of South Africa. This subspecies has the unique ability to recover from losing its queen because some of the laying workers have the ability to lay diploid eggs by a process called thelytoky. Without fertilisation, these eggs hatch, providing a source of worker larvae that can be fed to develop into queens.
>
> This sounds like an outstanding advantage doesn't it? Why haven't all the honey bee subspecies developed the ability to recover in this way?
>
> The answer is that it has a significant downside and this is the creation of competition. The workers that can lay diploid eggs are inevitably in fierce competition with each other as to which can be the mother of the new queen – who will be a clone of themselves. This results in a prolonged period of anarchy within the colony, during which queen cells are started and then torn down. During this time there is no laying queen and the colony dwindles.
>
> This trait is thought to have developed in response to the windy climate on the Cape in which queen loss during mating is an important factor. This trait can therefore be seen as a trade-off between the risk of queen loss during mating and the disruption caused by competition between individuals.

An extreme example of a mating swarm

In the course of the mating process, worker behaviour towards the queen changes from the indifference shown towards a virgin to the 'respect' and close attention given to a laying queen. This is because she gradually develops a mix of pheromones and the better mated she is, the more she attracts attention from the workers.

Even with an escort of workers to support and guide her, a queen often travels a considerable distance to visit a DCA, which is inherently a very risky strategy. If the queen does not make it back to her colony there is no backup queen and her loss means that the colony will inevitably die out.

When a virgin queen goes on a mating flight she usually leaves the colony in a state of excitement which manifests itself as bees crowding around the entrance and on the front wall of the hive. This behaviour is sometimes so exaggerated that it is referred to as a mating swarm. It persists until the queen returns 15–30 minutes later. If the beekeeper does not know the status of the colony it could easily be mistaken for the beginning or end of a true swarm.

After the queen has mated, workers pay her close attention

An attentive beekeeper, noting that re-queening has failed, can of course introduce a frame containing eggs and/or young larvae from another colony thus giving the colony a second chance at raising a queen – something that could never occur naturally. Most beekeepers regard this as a way of 'saving a colony' which, from their point of view, is true. However, for the colony itself this is (genetically) the final nail in the coffin because the remaining bees have now become 'slaves' to a non-related queen (who is a sort of cuckoo if you like).

As soon as the non-related queen comes into lay, the existing workers will quickly be replaced by her progeny and the original genetic line will be terminated (ie, they will have lost out in the race for survival). The idea that a honey bee colony is 'immortal' can thus be seen as ridiculous. Each new queen has her own unique genetic identity which, in a matter of a few weeks, will come to dominate the colony.

To enable a smooth transition, the new queen receives initial support from workers carrying her mother's genetic make-up. Helping their mother's daughter (their sister) in this way also enhances the old queen's success – it's a bit like the role of grandparents in a human family (but from personal experience it's more fun to pass on bad habits than accumulated wisdom!).

The advantages of polyandry

The advantages of the queen mating with numerous drones are many and seem to cover virtually all aspects of colony fitness. Compared with a colony where the queen has mated with a limited number of drones, one with a high genetic diversity produces more comb and more brood and collects more honey – it does everything better! It is also less prone to disease, better at thermoregulation and has a higher chance of long-term survival. Exactly how this all works is another matter; some things are quite well understood but others not. Perhaps rather surprisingly, genetically diverse colonies have been shown to have a lower varroa population. However, how a genetically diverse colony that produces more brood can have fewer varroa lacks a satisfactory explanation.

The downside of polyandry (apart from the risk of queen loss)

Every living organism wants to maximise the chance of passing its genes on to the next generation – and on into the future. This means that individuals are inevitably in competition with each other to pass on their genes. A honey bee colony normally has only one queen for this very reason, because if there were two or more they would be in competition with each other and harmonious functioning would be compromised (impossible?). This is why virgin queens fight and kill each other until there is only one left – or so it is said in many beekeeping books.

However, it is probably not as simple as 'last queen standing takes all'. There is plenty of evidence that the choice of a new queen is determined by a process slightly more sophisticated than sororicide (killing one's sister), in which the workers themselves play an important role.

A two-queen colony

We recently observed first hand what happens to a colony that has two queens; in this case one was at the bottom of the hive and one at the top with a shared storage area (of four supers) in between.

The queen at the top was the new one (daughter of the queen at the bottom) and when she had been in lay long enough to have her own family, it created a situation where two families of workers were each operating in the interest of their own mother whilst having to share a common storage area in the hive and the entrance at the bottom of the hive. This meant that some bees had actually to pass through the brood nest of a of a queen which was not their own mother.

At the time we were unaware of the situation and all we knew was that the colony gradually became more and more defensive and difficult to handle to the point where it had to be left as the last colony in the apiary to be inspected – and even then rather quickly before beating a hasty retreat.

This certainly was a problem for the beekeeper and must also have been for the bees because of the time wasted in defensive behaviour, but as far as we know it did at least fall short of civil war with bees killing each other.

The situation was not identified until the end of the season when the honey was harvested and, as soon as the two queens were separated (by the daughter queen being re-located), normal behaviour was resumed. This level of intolerance of more than one queen may be a feature of *Apis mellifera mellifera* which only rarely undergoes a perfect supersedure. Other races, eg, *A.m. ligustica*, seem to be more tolerant thus allowing a two-queen system to work harmoniously – or so it is claimed.

Some ant species have evolved to have several queens at the same time but only because these operate as sub-colonies (with spatial isolation) within a sort of super-colony, but in the end one queen often comes to dominate. Not so the honey bee colony, which can only operate smoothly with one queen, the one exception being during a perfect supersedure, when mother and daughter are able to coexist in apparent harmony, sometimes for quite a protracted period.

Kinship

A potential problem arises because honey bee drones are haploid and all the sperm that an individual drone produces have identical genetic composition. This means that the population of workers in a colony consists of a series of sisterships (the technical name is **patrilines**, *pater* being the Latin word for father).

Unlike normal siblings (where both parents are diploid) who have 50% of their genetic make-up in common, the individuals of a patriline (full sisters) share 75% of their genes (50% from their drone father and 25% from their mother). By contrast they only share 25% of their genes with their half-sisters (those that came from their common mother). The family relationships may be further complicated by the fact that a queen is likely to have mated with drones who are brothers (sons of the same queen) and this creates intermediate relationships in which patrilines share 50% of their genes.

Genetic relationships

As you can see, the honey bee has a rather odd sort of family according to our human way of looking at things – but what's the problem?

The problem arises because the workers are well aware of these genetic relationships. This is called kin recognition and is thought to result from subtle genetically determined

Worker bees in a colony can recognise their full- and half-sisters

differences in cuticular chemistry. Bees live in a chemical world (which is the source of much of the information on which they base their activity) and when interacting with other bees in the colony, they can discriminate between full- and half-sisters. This ability is known to extend to the larvae and pupae of both workers and queens! This should lead to **nepotism** (favouring one's close relatives) but in everyday colony activity this does not seem to manifest itself. It has been shown that nurse bees show a slight preference to feed larvae which are their own full sisters but not to the extent that half-sisters are neglected or that it affects the functioning of the colony.

One of the advantages of genetic diversity is that some genotypes provide the colony with specialised skills. It is also possible that kin recognition plays a part in the recruitment of bees to carry out specialised tasks (nepotistic recruitment?).

Despite the fact that polyandry and kin recognition have the potential to be disruptive (cause civil unrest) through nepotism, by some means the colony has strategies that keep this under control under most circumstances. However, in one aspect of colony behaviour, ie, during cast swarming, just a glimpse of nepotistic behaviour can be observed but in this context it is unlikely to cause any problems (to be maladaptive) at a colony level. This will be discussed again in Chapter 1.8.

1.3 HOW WE REACHED OUR CURRENT UNDERSTANDING OF SWARMING

The understanding of the process by which a colony of bees swarms is far from complete. The gaps in our knowledge do not exist because of lack of curiosity or scientific endeavour but reflect the complexity of a process that involves many thousands of bees and the fact that most of it takes place in the dark within the confines of the nest. Only the behaviour of the swarm once it has left the hive and is in the process of locating a new nest site is well understood and is beautifully explained in Tom Seeley's book *Honeybee Democracy*.

Most beekeeping books say little or nothing about the biology of swarming, apart from some discussion of the triggers for swarming and the bare essentials of the chronology of the process as seen by the beekeeper. These accounts also tend to be very plagiaristic, simply repeating what previous (often highly revered) authors have said in the past. There tends to be very little new input taking account of information that has come to light through bee science. This is the way in which beekeeping misconceptions and myths – of which there are many – are preserved.

SWARMING BIOLOGY AND CONTROL

Unreliable source of information

As we found in our early days of beekeeping, the testimony of other beekeepers is usually a rather unreliable source of information. They often prove to have little knowledge of the current condition of their colonies (whether they have swarmed or not) and what is likely to happen next.

The most honest description of the beekeeper's attitude to swarming appears in a book *Beekeeping: A Practical Guide* by the American author, Richard E Bonney. In the introduction to his chapter on swarming he writes, *'For some swarming is inevitable and there is no way to stop it. For others failure to control swarming is seen as a measure of beekeeping competence, in themselves or others. A competent beekeeper would never allow swarming, so they think.'*

There are still beekeepers out there who think that the only method of swarm control is to destroy queen cells with no thought as to the condition of the colony and why the cells are there in the first place.

Destroying queen cells is not a method of swarm control

More reliable knowledge

Bee science provides more up-to-date and reliable knowledge but, because some bee scientists have rather limited practical beekeeping experience, the interpretation of their findings and how it applies to the management of colonies is sometimes lacking.

Some parts of the swarming process have been little investigated because of the problems involved. Studies using observation hives have been able to elucidate many of the behaviours involved in swarming but the small size of the colony being studied creates some doubt as to the validity of all of the findings.

Unfortunately cast swarming (the issue of after-swarms) has attracted very little attention and seems to have been passed off as 'just another swarm', but from a practical point of view it is quite important. Little thought has been given as to why bee colonies produce cast swarms and how they are organised but in many ways they are different from the prime swarm (not just by being headed by a virgin queen) and this is further discussed in Chapters 1.7 and 1.8.

Practical experience

The other source of information, and possibly the most important one (in my personal view), is practical beekeeping; what the beekeeper can see during the day-to-day management of colonies.

Opening a hive and finding it at some stage in the process of swarming provides a snapshot of what is going on. With careful observation these snapshots can be interpreted rather like time-lapse photography, albeit with rather large time intervals. Some behaviours cannot be observed at all because the very act of opening a hive disturbs what is going on inside. However, the individual observations can be used rather like those puzzles that appear in children's books where numbered points have to be joined up to reveal the underlying picture.

Working hypotheses

In the first half of this book it is my intention to try and give a complete (beginning to end) biological account of swarming, including the much neglected subject of cast swarming.

Because there is so little known about some parts of the process, much of this will take the form of a **working hypothesis**. This is a framework which incorporates information from a wide range of sources (books, science and practical beekeeping), that can be used to guide the beekeeper's interventions with the colony (management) so that they do not conflict with what is the most important part of the colony's life cycle (reproduction), whilst at the same time preventing loss of bees and, with them, a significant part of the honey crop.

In the absence of a better understanding, a working hypothesis provides useful guidance and remains valid until exceptions (failures or contradictions) arise, when it may have to be

> **An example of a working hypothesis**
>
> A good example of a working hypothesis is 'what goes up must come down'. So, if you throw an object upwards it will not continue in that direction indefinitely but will eventually fall back down. If it was a heavy object then application of the theory says that it is unwise to be standing underneath when that happens. So this is quite a useful working hypothesis on the whole!
>
> It wasn't until Isaac Newton came on the scene and propounded the theory of gravity that it was better understood and no contradictions were revealed. I used the word 'better' because the force of gravity is still not fully understood today but that does not mean that things are going to start falling upwards anytime soon – so the hypothesis is still working!

HOW WE REACHED OUR CURRENT UNDERSTANDING OF SWARMING

modified or even abandoned (ie, it's only useful as long as it works!). When referring to a hypothesis I will try to make its status clear and also provide such supporting evidence that is available at the present time.

SWARMING BIOLOGY AND CONTROL

1.4 THE THREE TYPES OF QUEEN CELL – THEIR ORIGIN AND FUNCTION

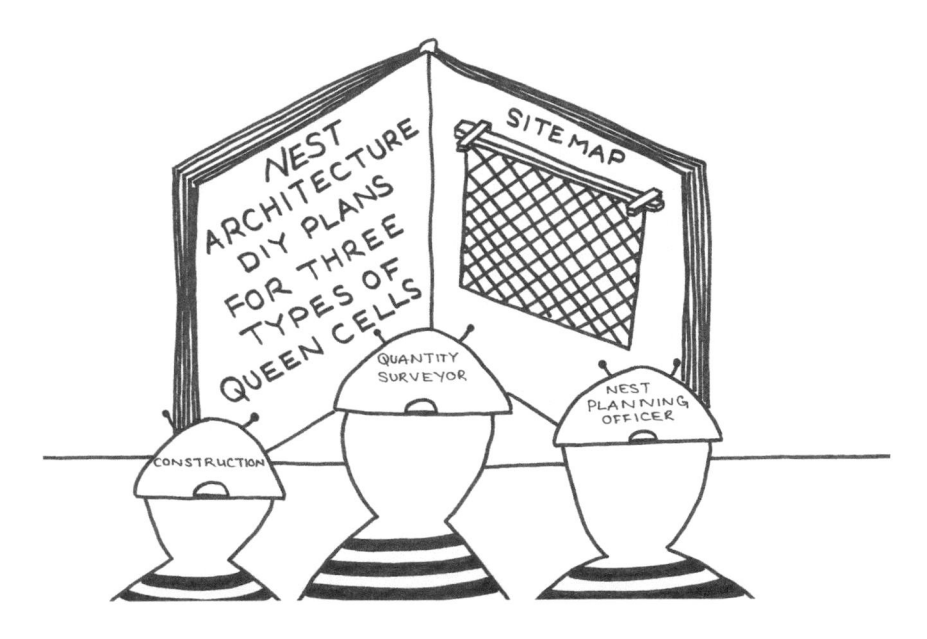

When a beekeeper opens a hive and sees queen cells it should not be automatically assumed that the colony is going to use these to swarm because they may be there for an entirely different reason.

Reasons for queen cells being present in a hive

There are three different reasons why a colony produces queen cells:

- Reproduction – swarming
- Replacement of the existing queen – supersedure
- The colony has lost its queen (it is queen-less) – emergency re-queening.

Let there be no misunderstanding, these are three separate and distinct behavioural programmes that the colony is following, but nine times out of ten (or probably more), when

SWARMING BIOLOGY AND CONTROL

a beekeeper opens a hive and sees queen cells, it is the **swarming programme** that the bees have embarked upon.

Different types of queen cells

We often refer to the 'types' of queen cell because the origin of the cell and its contents, its position in the colony and the number of cells that have been produced are usually characteristic of the programme that has been activated. These are all important clues as to what is going on but are not completely definitive (see 'Ambiguous situations' below). I have even heard a beekeeper claim that a colony had all three types of queen cells present at the same time, but this is meaningless because the programmes are mutually exclusive.

Before deciding on any management in response to finding queen cells it is necessary to correctly identify which programme the colony is following. Figures 1–3 show typical queen cells (their external appearance and position on the frame) for the three programmes. Only the presence of **swarm cells** means that the colony is intent on swarming and those produced for other reasons (supersedure and emergency re-queening) will on **NO** account result in the issue of a swarm. The presence of supersedure and emergency queen cells does not usually require any intervention from the beekeeper – except to **leave the bees well alone and let the colony get on with it**.

So, how do you know which programme has been activated?

1 Swarm cells

These are developed from queen cups into which eggs have been laid by the queen (or transferred by workers) and are entirely vertical. The cells are long and usually constructed on the edges of the combs; either along the bottom bars or in recesses by the sidebars of the frame (see Figure 1). The longest cells are usually founded on a boss of wax which

THE THREE TYPES OF QUEEN CELL

Figure 1
Swarm cells in a typical position

Figure 2
A supersedure cell (queen emerged)

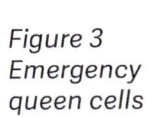

Figure 3
Emergency queen cells

53

enhances their apparent size but the internal volume of queen cells (which is what really matters) is very uniform. Occasionally some swarm cells will be on the face of a comb – particularly when the brood nest is in a single box.

In terms of number, there are rarely less than 5–6 queen cells, more typically 10–20 and possibly up to 100. As the name suggests, the colony is producing new queens so that it can swarm and unless the beekeeper intervenes, in due course, this is exactly what will happen.

The issue of a swarm usually occurs around the time that the first queen cells are sealed (Day 8) but it can occur earlier, especially if the beekeeper has previously destroyed queen cells. Swarming can also occur early for no apparent reason – some colonies just do! It can also occur later if it is delayed by poor weather and, in extreme cases, virgin queens may be ready to emerge from their cells but 'warder' bees keep them imprisoned until their mother has departed with the prime swarm.

2 Supersedure cells

Like swarm cells, these are typically vertical, usually located on the face of the comb and develop from an egg laid in a queen cup by the queen (see Figure 2). But they can also have the same origin as emergency queen cells (see below) being based on an egg laid in a worker cell (not a queen cup) – they look like an emergency cell but have been constructed for a quite different purpose.

In supersedure there are usually only two or three cells grouped together on the same frame. The intention here is to replace the existing queen who, for whatever reason, the bees have decided is no longer up to the job. She may be old, she may be damaged or probably a host of other things of which we are not aware. Unfortunately one of the queen's defects that the bees often seem unable to detect is when she is running out of sperm and is destined to become a drone-layer.

The books say that during supersedure the old queen is normally retained until her replacement has mated successfully and started to lay.

In some cases, mother and daughter coexist in the hive for some time but eventually the old queen will 'disappear'. This is called a 'perfect supersedure' but with the type of bees we have in Wales 'perfection' seems to be the exception rather than the rule. More usually the supersedure is 'imperfect' and there is a brood gap of a week or more as a result of the old queen having been eliminated before her daughter starts to lay.

It is not known who 'assassinates' the old queen; whether it is the new queen or the workers or it is a co-operative activity. If supersedure cells are found in a colony there is nothing the beekeeper needs to do except **leave well alone and hope the outcome will be successful**.

Early spring and late autumn attempts at supersedure are often unsuccessful – usually because of a lack of drones. Later in the season there is also an increased probability of the queen becoming infected with deformed wing virus (DWV) during mating which may result in her early failure. Early and late supersedures both need careful monitoring to ensure the colony does not become queen-less but, unless a spare mated queen is available, nothing can be done about it.

Because hive inspections are less frequent, in the autumn, late-season supersedure may go unnoticed. When making an early spring inspection, a beekeeper can often be surprised to find that the colony does not have the same queen that was there in the late summer. Unbeknown to the beekeeper, the old queen has been superseded and her replacement is unmarked.

3 Emergency queen cells

These are queen cells produced in response to the sudden loss of the queen. Their identity is unambiguous because there will have been no queen in the colony since the day when the cells were started and the younger stages of brood (and particularly eggs) will be missing.

As the name implies, emergency queen cells are produced in response to the worst possible situation in which a colony can find itself (with no queen) and its only aim is to get a new queen as soon as possible. **The last thing such a colony wants to do is to swarm and further jeopardise its survival.**

The queen may have died suddenly of natural causes or the beekeeper may have killed her or spilt her onto the ground during colony manipulations. Emergency queen cells are also produced if the beekeeper deliberately removes the queen from a colony.

Occasionally a colony may lose its queen more than six days after she last laid any eggs and, in this circumstance, there will be no brood young enough to make an emergency queen as larvae three or more days old cannot be raised as queens. Without intervention by the beekeeper this colony will be unable to re-queen itself and will die out.

There are usually numerous emergency queen cells (see Figure 3) but instead of being developed from an egg or larva in a queen cup, they are based on existing eggs or young larvae in normal horizontal worker cells. Nurse bees start to feed the selected occupant with royal jelly and the outer rim of the cell is extended downwards to make room for the increasing size of a queen larva.

There seem to be two types of emergency queen cell: one type has the surrounding comb extensively modified to produce a vertical cell (very similar to a swarm cell but inset on the comb face – see Figure 4a) and the other type is part horizontal and part vertical with a right-angle bend in the middle (see Figure 4b). Although a bent cell (externally) looks inferior, both types seem to contain queens that are indistinguishable in terms of size.

THE THREE TYPES OF QUEEN CELL

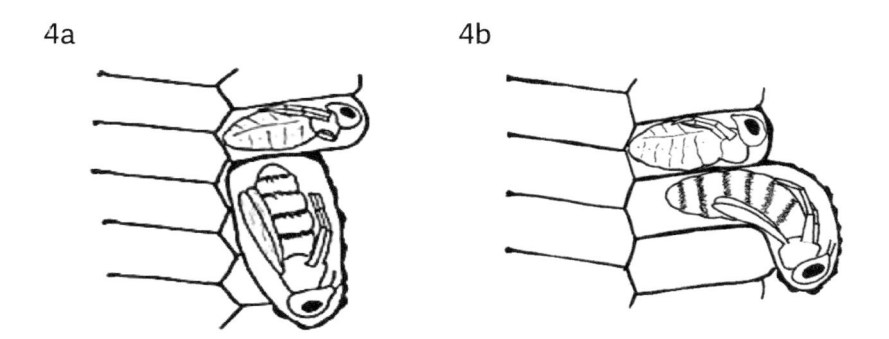

4a   4b

Figure 4a   An emergency queen cell where the surrounding comb has been modified to produce a vertical cell
Figure 4b   An emergency queen cell built directly from the cells on the comb

The fully vertical cell is most common on new comb and the right-angled version on old comb – presumably because of the difficulty of nibbling away cell walls that are reinforced by several layers of tough pupal skins.

At first sight, emergency queen cells look rather unimpressive when compared to swarm cells. They can also be easily overlooked unless the bees are shaken or brushed from the comb.

Inferior (scrub) queens

There is a firmly rooted myth in beekeeping that queens developed in emergency re-queening are inferior to those produced during swarming. Despite looking smaller than swarm cells from the outside, emergency cells normally produce perfectly good queens with a full complement of ovarioles (egg-producing units).

Recent work on emergency queen cells has revealed that the bees carefully select larvae that already have good nutritional status for promotion to queens. It is extremely rare for them to select a larva that is too old – providing ones of the correct age are available.

The idea that emergency queen cells produce an inferior queen (a 'scrub' queen) is probably based on last-ditch attempts by beekeepers to re-queen colonies that have been queen-less for some time by giving them a frame with eggs or young larvae on it. Such colonies that are populated by OAPs simply do not have enough nurse bees of the right age to produce a fully developed queen. Older bees, and particularly those that have started to forage, have atrophied hypopharyngeal (brood food) glands and, although they can rejuvenate them to some extent, it takes time and time is of the essence in this situation.

Ambiguous situations

In the vast majority of cases, when a beekeeper opens a hive and finds a significant number of queen cells the intentions of the colony are completely obvious – it is **SWARMING** and there is no room for confusion. However, it should be emphasised that it is not the appearance of queen cells or their position on the frame that is most important, it is why they are there in the first place; in other words, the behavioural programme that the colony is following.

Swarming or supersedure?

Doubts sometimes arise when there is a small number of queen cells – is it swarming or supersedure? This is not always obvious. For example, swarm cells are not always at the edges of frames and supersedure cells are not always on the face of the comb and, just to confuse matters, the number of cells may be atypical – too few to be typical of swarming or too many to be supersedure? Fortunately emergency re-queening is always obvious because there will be no eggs in the hive and the youngest brood will tell you exactly when the queen was lost (together with the age of the queen cells).

How to deal with these situations, where what the colony is doing is in doubt, is dealt with in more detail in Chapter 2.3.

THE THREE TYPES OF QUEEN CELL

The number of queen cells

Recent observations have shown that in some years (particularly early in the season) only a small number of (swarm) queen cells are produced and these are usually in a peripheral part of the brood nest (not on central combs). This can be somewhat confusing. At other times large numbers of swarm cells are produced and there is absolutely no mistaking that the colony intends to swarm. A possible explanation for the phenomenon is discussed in Chapter 1.5.

Swarm and emergency cells together

There are some circumstances where swarm and emergency type (both origin and configuration) queen cells can coexist in a colony. For example, if a colony swarms early (before the swarm cells are sealed), the workers will often respond to what they perceive as the loss of the queen by making some additional emergency cells. This is a direct response to the sudden loss of queen pheromones in the colony and the bees that are responsible seem to be unaware that they already have numerous swarm cells present.

The same thing can happen if the beekeeper removes the queen when making an artificial swarm – the colony already has more than enough queen cells but nevertheless makes more.

In neither of these cases has the colony changed to an emergency re-queening programme and the emergency cells are usually of no practical significance; they are so much younger than the swarm cells that they are unlikely to survive to maturity.

When emergency cells are significant

However, there is one situation where emergency queen cells can matter and that is when a colony has already swarmed and the beekeeper has been forced into destroying all but one of the remaining queen cells to prevent the issue of a cast swarm.

If the culling is done immediately after the prime swarm has departed, there will still be eggs and young larvae present from which the bees can make emergency queen cells. These cells will be started after the beekeeper has closed up the hive, satisfied that everything is under control.

Being in swarming mode, the colony may proceed to cast swarm with the queen that emerges from the queen cell the beekeeper carefully selected and left intact, treating the emergency cells as backup to provide their new queen. In this way they have outwitted the beekeeper and if bees could laugh this would be an opportunity to do so at the expense of the beekeeper.

Again, it is not the type of cell that matters but the behavioural programme the colony is on that determines the outcome – and the management that is required to deal with this particular situation can be found in Chapter 2.3, Steps 5 and 7.

Understand the bees' programme

No management should be applied to a colony that has queen cells until it understood which programme the bees are following. If it is found to be swarming, then the same rule applies; nothing should be done until it has been determined what stage in the swarming process the colony has reached.

Chapter 2.3 contains a 12-step diagnostic key to help the beekeeper to make an accurate diagnosis.

1.5 TRIGGERS FOR SWARMING AND THE START OF QUEEN CELLS

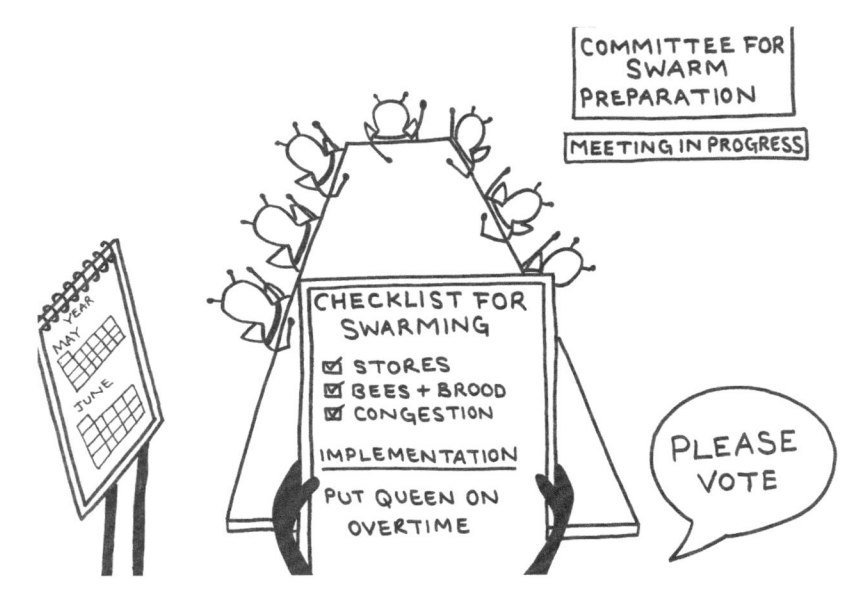

How does a honey bee colony know when to swarm?

Over the years there has been much discussion about this topic in beekeeping books and magazines and every so often someone claims that they have 'discovered the secret of swarming'.

In real life the triggers for swarming are almost certainly multifactorial: a function of colony size, brood area, space for the queen to lay, the adult population and its age distribution (a large population with a low mean age), brood nest congestion (the number of bees crammed into the available space) and the age of the queen. External factors also seem to have some part to play: the weather, hive temperature and ventilation, nectar and pollen availability and day length – take your pick!

Choosing the right time to swarm is a tricky decision for a bee colony and can make the difference between success and failure in this its most important function – to reproduce.

Preferred cavity size

In the wild, colonies have strict criteria for the size of cavity in which they choose to establish their nest and this has a preferred volume of about 40 litres, regardless of the shape. To give some idea of what this means, a Modified National deep box has a volume of about 35 litres.

The natural nest size is chosen to provide the optimum space in which the colony can complete its entire life cycle (and that includes enough space for storage of honey to enable it to survive the winter). During the spring build-up the colony seeks to fill its nest cavity with bees, brood and stores as early in the season as possible. When this task has been accomplished, the colony regards itself as being in a position to swarm.

After swarming (and probably cast swarming), establishing a new queen in the old nest site, rebuilding the colony size, producing winter bees and accumulating sufficient stores for the oncoming winter become the priorities.

A natural nest cavity in a tree

TRIGGERS FOR SWARMING AND THE START OF QUEEN CELLS

Colonies with some degree of hybridisation, usually with Italian bees, can exhibit a second swarming period later in the year

In most seasons, early swarming gives the best chance for the swarm itself to find a new home and to become mature and fully established before winter – and for the parent colony to establish a new queen and recover.

Second swarming period

Races of bee adapted to a Mediterranean-type climate have a second period in which they can potentially set up to swarm and this is in the late summer or early autumn.

In this climatic regime, the main flowering period is in the spring and this is followed by a hot, dry summer when there is little available forage. When the autumn rains arrive, there is a second flowering period which may be sufficient to induce the colony to make a second attempt at reproduction.

Cold-temperate-adapted races do not have this behaviour and if you encounter a swarm after July in the UK, in my experience anyway, it is because it has some degree of hybridisation, usually with Italian bees (*Apis mellifera ligustica*), ie, the bees and/or the queen have a distinctly yellow 'tinge'.

Adjustable space

Returning now to the matter of nest volume, keeping our bees in modular hives in which the volume can readily be adjusted is pushing the boundaries of evolved behaviour and the question is, how much does this matter? So-called 'natural beekeepers' would claim that we should not interfere in the swarming process – and they have a point.

One thing management of space does is to at least prevent early swarming. However, if and when the colony swarms, this delay has implications for swarm survival. Even though a later swarm is usually larger, the lateness of the season gives it less time to establish itself in a new nest site and for the parent colony to re-queen and rebuild. Swarming is the main source of new or replacement queens so if the colony is prevented from swarming it has some implications – maybe not immediately but at least in the long term.

By catching and hiving a late swarm and, by other means, ensuring the colony gets a new queen when it needs one, the disadvantages (for the colony) of late swarming can be minimised. There are other issues that arise as a result of practising swarm control, such as the choice of the new queen (who does it, the colony or the beekeeper?) that are probably more important (see Chapter 1.8).

External conditions

Although hive management can have a major influence on internal (hive) conditions, the beekeeper has no control whatsoever over external conditions – except possibly through the choice of apiary site.

The main trigger for swarming is thought to be the production and distribution of queen substance. As the queen ages

Congestion in the hive contributes to swarming

she produces less queen substance and this is supposed to predispose a colony with an older queen to swarm. Congestion is said to promote swarming as a result of poor distribution of the queen's pheromones within the colony.

Supplementing the supply of queen substance in a colony has been suggested as a means of preventing swarming. However, experiments in which synthetic queen substance was added to hives have not proved successful in preventing colonies from swarming. So either queen substance is not the sole trigger or the synthetic version was not a correct substitute for the real thing. The active ingredient of queen substance in relation to swarming is thought to be 9-oxo-(E)-2-decenoic acid but there are several others that could be involved in this particular colony response.

Weather is probably the least considered trigger for swarming and is rarely mentioned in beekeeping books. The exception is LE Snelgrove who, writing about swarming, noted, 'In our part of the world [he lived in north Somerset] the vagaries of the weather present unexpected problems to the apiarist which tax to the utmost his patience and resourcefulness'. Poor weather limits flying time so bees stay home and cram into the brood area, either because they have no other work to do or to keep themselves and the brood warm. This often simulates congestion and triggers the production of queen cells.

Conditions below the queen excluder are more important than those above. In adverse weather, even generous supering will not prevent the bees from cramming themselves into the lower part of the hive, thereby creating congestion. Typically, a colony perceives itself as ready to swarm when it is at the peak of its seasonal build-up, at which time up to half of the workers can be less than eight days old. However, it is unlikely that this is something that just happens by chance and there is evidence that it is part of a co-ordinated plan by the bees to prepare themselves for what, to them, is the most important event in their annual life cycle, the day when they can swarm (see Chapter 1.6).

1.6 THE START OF QUEEN CELLS TO THE ISSUE OF THE PRIME SWARM

SWARMING FOR DUMMIES

CHAPTERS:
1. HOW TO START QUEEN CELLS
2. FEEDING LARVAE-DIET
3. QUEEN ON DIET AND EXERCISE
4. EXTENDING QUEEN CELLS
5. COUNTDOWN TO SWARMING
6. CHOOSING WHEN TO SWARM
7. LOAD HONEY
8. TEAM SELECTION
9. WARM UP PERIOD
10. START TO SWARM

The sequence of events over this period is well known, but how it is organised and by whom is poorly understood. Maturation of the queen cells is the most obvious development; the continued feeding of the queen larvae by nurse bees and the extension of the cell walls until they are sealed on Day 8 after the laying of the egg (or Day 5 after the larva hatched and started to be fed royal jelly). This is followed by the issue of a swarm (usually) soon after the queen cells have been sealed.

The previous chapter discussed the triggers for swarming and these can be summarised as being multifactorial, ie, there is a wide range of sources of information (both internal and external) that a colony of bees can utilise to make the decision to start the swarming process. Once it is started it is rarely abandoned – at least in response to natural events.

Some beekeepers claim that the onset of an abundant nectar flow can cause bees to abandon the idea of swarming but I have never known this to happen and I don't know of any hard evidence that this is true. Radical brood nest management, such as the removal of a substantial number of frames of brood and their replacement with foundation (simulating brood nest immaturity), occasionally works but cannot be relied on.

From a practical beekeeping perspective, when queen cells are produced with the intention of swarming, it is best to regard the outcome as inevitable, ie, the process will run full term unless 'diverted' by management.

At what point does the intention to swarm start?

Beekeeping books have cited a number of stages in the development of a colony as the start of the swarming process:

- drone brood in the colony
- adult drones in the colony
- queen cups constructed
- eggs laid in queen cups
- queen cups containing young larvae lying in pools of royal jelly.

The first three of these stages can be found in colonies that may never (during that season) progress further and attempt to swarm.

The fourth one (eggs in queen cups) is more difficult to interpret; is this a serious warning or not? If a close watch is kept on queen cups, eggs can often be observed to come and go without being in place long enough to hatch and start to be fed with royal jelly. These eggs are almost invariably standing upright in the cell, meaning that they have been laid within the past 24 hours. The fact that they disappear presumably means they have been deliberately removed by the workers but it is not known why.

THE START OF QUEEN CELLS TO THE ISSUE OF THE PRIME SWARM

Drone brood in the colony

Adult drones in the colony

A queen cup

An egg laid in a queen cup

A queen cup containing a young larva lying in a pool of royal jelly

The question is, who laid these eggs? Is it a premature (false) start by the queen, acting before the rest of the colony is in agreement? It could hardly be an accident because the queen should realise what she is doing by the orientation and size of the cell which she will have assessed before laying. The diameter of queen cups is roughly the same as that of drone cells so, unless they have been chemically marked in some way, orientation must be her guide.

Laying workers?

Another possibility is that these short-lived eggs are worker-laid. It is known that virtually all colonies have a few workers that lay a small number of eggs (about 1% of the total – but unfertilised, of course). These do not have the queen's chemical marking and so could be quickly recognised and removed by the 'worker police'. Worker eggs would more commonly be laid in both worker and drone cells but, in these locations, the beekeeper would be unable to distinguish them from queen-laid eggs so we are not aware of their presence or subsequent removal.

So it is possible that finding (upright) eggs in queen cups is nothing whatsoever to do with swarming and should not be taken seriously. No definitive answer is possible at our present state of knowledge.

Finding young larvae in pools of royal jelly is usually regarded as a sure sign that the colony is triggered to swarm. Very occasionally the process seems to self-abort – for reasons that are not obvious. As already noted, radical brood nest management can stop the process but too rarely to afford a practical solution. Destroying queen cells (even repeatedly) does not usually prevent swarming. At best it may delay it by a few days and at worst it can result in rendering the colony queen-less. So when queen cells have been started it is best to regard swarming as inevitable.

THE START OF QUEEN CELLS TO THE ISSUE OF THE PRIME SWARM

Colony unanimity

After closely monitoring colonies for the start of queen cells over a number of years, it has become obvious that the number produced varies quite widely; from fewer than 5 up to 50 or more. This is usually regarded as being a characteristic of the colony itself and colonies that produce a large number of queen cells should be regarded as 'swarmy'.

Some colonies produce a lot of queen cells!

However, it does not appear to be as simple as this because there are seasons (or parts of seasons) in which it has been observed that virtually all colonies produce only a small number of queen cells. In other seasons they will go to the other extreme and all tend to produce a lot – and there are still other seasons where moderation is the name of the game. Other beekeepers confirm this variation in queen cell numbers and it often seems to be synchronised with what is happening in other colonies.

The question is, what causes this variation?

It is possible that pre-emptive swarm control may be implicated. Management to ensure that the queen always has somewhere to lay and there is no shortage of space for the storage and processing of nectar almost certainly serves to weaken the triggers for swarming – it's like 'moving the goalposts'. In some seasons it may fail to prevent any attempt to swarm but it will always serve to delay it.

Lack of clear-cut information as to whether or not now is the right time to swarm, resulting in a lack of colony unanimity, may be a root cause of the variation in the number of queen cells. So it is an interesting question. Is rigorous pre-emptive swarm control a bad thing for the colony?

If completely successful for several years (which it rarely is), it may prevent the colony from getting a new queen and, if an aging queen is not superseded, she may fail the colony at a time when they can do nothing about it. Also, from the beekeeper's point of view, an older queen may cause the colony to be less productive. This is discussed further in Chapter 2.2.

'Swarm control' bees

Later in this chapter it is hypothesised that the swarming process is not a 'whole colony matter' but is both initiated and controlled by a relatively small group of bees – meanwhile the rest of the colony is getting on with its daily business in 'blissful ignorance'. We cannot readily identify this group of bees, except possibly at the time when swarming actually occurs.

Having checked that there are sealed queen cells and weather conditions are favourable, these are the bees that decide the time for the swarm to issue and gee-up the colony in preparation for the great event. It is not known whether it is the same group of bees throughout the process (which, as we will see in a minute, lasts more than just eight days) or whether the responsibility is passed on to other bees – but the former appears more likely. The partitioning of responsibility to a relatively small group of self-appointed bees which identifies what needs to be done and proceeds to do it without much reference to the rest of the colony, seems to be the way in which some of the more specialised activities within the colony operate.

Decisions – how are they made and who makes them?

We have already discussed the multiple triggers that lead to swarming but who assesses these and comes to the decision that now is the right time? This must surely be done by a group of bees working together, gathering the necessary information on which to base the decision. But what if there is no unanimity within this group? How many bees does it take to constitute a quorum which is capable of proceeding to implementation? Can the controlling group be supplemented along the way?

It is already known this can happen when a swarm is trying to find a new home; the original scouts 'retire', and new ones take over the task. This is rather like the human activity of setting up an independent enquiry in which a new set of bees reassesses the available information on potential nest sites. Occasionally it is obvious that unanimity has not been reached and the swarm will set off in different directions but, because only one part contains the queen, the bees are forced to regroup and 'reconsider their verdict' (see Tom Seeley's book, *Honeybee Democracy*).

Following this line of argument, the variation in queen cell numbers could be caused simply by the number of bees that are 'in favour' of swarming. Occasionally colonies will start queen cells and then apparently give up on the idea and the cells will be torn down. This may be because the number of bees that initiated the process is too small and they have been 'out voted' by other bees which are not in favour. Or it could be that other issues have intervened and their attention has been diverted.

Positive decision

This must, I think, be a positive decision that is made before larvae in queen cups start to be fed royal jelly because, once started, queen cells are automatically fed by the nurse bees. Hence the ability of the beekeeper to place started queen cells in a queen-right colony and they will continue to be fed (on autopilot) until they are approaching maturity.

Small numbers of queen cells

Another characteristic of a colony that has produced only a small number of queen cells is that they are usually produced in peripheral (out-of-the-way) positions – almost as though this was covert activity (hoping that nobody would notice them there!). On seeing this situation (perhaps a marginal decision by the colony?) the beekeeper has three options:

- To treat the colony as though it was intent on swarming and do an artificial swarm.
- If the cells are some time off sealing, the beekeeper can wait a few days to see if more cells are produced before making a decision as to what action to take. If more cells are produced, revert to option 1.
- The cells can be destroyed in the hope this will prevent or delay swarming. But before doing this it must be verified that the queen is still in residence and there are eggs or young larvae from which replacement queens can be raised. If the colony is intent on swarming, more cells will be started pdq and this is an indication to revert to option 1.

THE START OF QUEEN CELLS TO THE ISSUE OF THE PRIME SWARM

We have found that queen cell destruction in these circumstances does often work and creates a worthwhile delay in swarming but it is rarely the ultimate (whole season) solution. When this is successful, what we don't know is, would the attempt to swarm have been aborted if we had not intervened?

WARNING – queen cell destruction needs to be used with great caution!

The point of no return

But is a larva in a pool of royal jelly the real start of the swarming process?

It certainly seems to be the point of no return.

Until fairly recently, I would have said that the decision to swarm occurred when newly hatched larvae in queen cups were given their first feed of royal jelly. However, studies carried out at the University of North Carolina, using observation hives to try and understand tactile communication (vibration signals and dances) within the colony, suggest otherwise.

Use of the vibration signal

Commencing about 14 days before queen cells are produced, and seemingly related to increased congestion in the brood nest, worker behaviour towards the queen changes. Starting gradually, but increasing in frequency with time, the queen becomes the target of what is called the 'vibration signal'. The worker giving this signal grasps the recipient (in this case the queen) between her front legs and vibrates her own body up and down for 1–2 seconds. It seems to be a multipurpose signal because it can be also executed on workers (of all ages), virgin queens and even queen cells.

Its meaning seems to be non-specific and is interpreted as being a generalised gee-up (stimulatory) signal, the response to which depends on the context in which it is delivered. In the period leading up to the production of queen cells, the vibratory signal, allied with increased feeding of the queen, means that she is stimulated to lay more eggs. Presumably the aim is to maximise the amount of sealed brood present in the colony at the point of swarming. The signal ceases abruptly just before queen cells are started.

> **Maximising resources**
>
> When later we come to consider the age-class composition of the bees in a prime swarm, this stimulation of the queen to lay more can be seen as advanced preparation to maximise the resource of young bees that can join the swarm.
>
> It also ensures that there is a reserve of sealed brood that will emerge only after the prime swarm has departed and these bees will be of the optimum age to support the new queen.

Immediately after the queen cells have been started, the vibration signal starts up again and the queen is subjected to it with increasing frequency up until just before the prime swarm issues. This time the vibration signal is allied with reduced feeding and has the aim of limiting the queen's egg production and forcing her to take more exercise, thus slimming her down in preparation for flight.

Egg laying to the last minute

Although it is usually claimed that queens have almost ceased laying completely by the time the swarm issues, careful inspection of colonies at this time shows that this is not always the case and there is wide variation in this aspect of queen behaviour. Some colonies have been found to contain large numbers of eggs that have actually been laid on the day of swarming. Interestingly, this trait seems to be associated with the production of a low number of queen cells so may also be the result of a less-than-unanimous decision to swarm.

Initiation of queen cells

But who initiates the start of queen cells? Is it the same bees that have been delivering the vibration signal to the queen during the previous couple of weeks? This is not known but a prime suspect must be the so-called 'messenger bees'.

In a stable colony, this is a role (an activity) undertaken by a small number of nurse bees up to 10 days old. For a period of a few days they show a strong attraction to queen substance. They can easily locate the queen by following her footprint pheromone on the comb and they form part of the queen's retinue for a short time. They are not there to tend her (feed and clean) but only to lick her body to obtain the mixture of pheromones, collectively known as queen substance.

Messenger bees in the queen's retinue are there to lick her body to pick up queen substance

The messenger bees then quickly disperse throughout the colony, making numerous contacts with other bees using their antennae and through trophallaxis. Their presumed function is to provide continuous confirmation to the colony that it is queen-right. Failure to get their message across with sufficient strength (possibly because of congestion) may be what kick-starts the swarming process. The question is, do these relatively young bees have sufficient knowledge of the state of the colony as a whole to make such a crucial decision themselves? Or does some other group/age-class of bees make this decision? As far as I am aware this not known.

In summary, this early change in behaviour and its continuity strongly suggest that the decision to swarm is actually taken about two weeks before queen cells are seen in a colony. If it is not actually a final decision to swarm, it is at least a decision to make advanced preparations. This is bad news for beekeepers because we (apparently) have no way of knowing when this decision has been taken. It means that we may often be applying pre-emptive swarm control management (such as adding frames of foundation) to colonies that are already triggered to swarm, unaware that there is little chance of success.

So in practical terms is there anything we can do to switch off what could be termed 'latent swarming'? All I can say is that when the brood nest is split fairly radically (either to provide pre-emptive swarm control or to make increase), the decision to swarm seems to be stopped in its tracks. This can be inferred from the fact that, when a colony has been subject to this type of management before it has started queen cells, the queen-right part of the split never sets up to swarm (at least within the next 3–4 weeks).

How do the eggs get into the queen cups in the first place?

The majority opinion is that they are laid there by the queen herself but what is the trigger for her to do this? One source suggests that it is the result of feedback from the messenger bees but, bearing in mind that their main focus is collecting queen substance from the queen, this sounds a bit unlikely.

Worker bees have been observed to make an intensive examination of queen cups, followed immediately by straddling them, exposing their Nasonov gland and fanning, presumably to attract the queen. The highly volatile Nasonov pheromone is usually used outside the hive to call in worker bees. When used inside the hive it would quickly be distributed everywhere.

So, does the queen herself respond to it or does it cause workers to escort the queen to where her services are required?

Others claim that workers transfer eggs (laid elsewhere) to queen cups. The two mechanisms are not mutually exclusive and whether they are queen-laid or transferred by a worker is of little practical significance as long as eggs get where they are wanted.

Special eggs

One recent piece of research has shown that the eggs a queen lays in queen cups are special; they are on average 13% heavier than those they lay in worker cells. This has been shown to confer advantages that persist throughout the developmental stages of the new queen which, on emergence, is slightly larger and also has more ovarioles than one that has been raised from an egg laid in a worker cell. Differences in gene switching have also been detected, involving both the production of pheromones and functioning of the queen's immune system.

It is not known if these differences are of practical significance but it does suggest that grafting of larva developed from eggs laid in worker cells may have some disadvantages. Whether the selection of larvae of high nutritional status to make emergency queen cells (see Chapter 1.3) confers some of the benefits of eggs laid in queen cups is not known.

Who decides that an egg in a queen cup should be allowed to hatch and who does the initial feed of royal jelly?

Again, it may be the messenger bees, or other nurse bees which are triggered by the messenger bees, but again this is not known. An allied question is, how are queen cells marked as queen cells so that the nurse bees have the correct dietary information? Is it the orientation of the cell or does the queen mark them differently when she lays the egg?

After a larva has passed the point where it has started to transform into a queen, the pheromones that it produces will provide the nurse bees with the information they need. However, this system would not work for emergency queen cells which obviously cannot be specially marked by the queen. Also, at least initially, an emergency queen cell does not have the correct orientation (it is not vertical) and has to be progressively modified. So are workers capable of marking cells containing larvae that are destined to be raised as queens? Again this is not known.

Once the initial feed of royal jelly has been given to an intended queen larva, all subsequent feeding seems to proceed on autopilot. This is why, in conventional queen rearing, grafted queen cells are transferred from a queen-less 'starter colony' (which does the initial feed following the emergency re-queening programme) to a queen-right 'finisher colony', where feeding is completed.

Despite the colony being queen-right, the cells continue to be fed (no questions asked) until such time as the virgin queens are nearing maturity, when (if they are not harvested or protected) the cells will be torn down by the workers and the occupants killed. Some queen breeders do not bother with a finisher colony but simply leave the queen cells in the starter colony but this has to be constantly 'replenished' with new nurse bees which is done by adding frames of sealed brood from donor colonies every few days.

THE START OF QUEEN CELLS TO THE ISSUE OF THE PRIME SWARM

When the pupating queens are nearing maturity, the cells are torn down by the workers and the occupants killed

Which bees in the colony 'know' they are going to swarm and make advanced preparation for S-day?

Who 'tells' them and at what stage in the process do they 'know'? Is it done on a 'need-to-know' basis – is it progressive?

There are no clear answers to these questions but it should be emphasised that up to 50% of the workers in a colony at the point of swarming had not even emerged when the queen cells were started. Nevertheless, many of these very young bees will issue with the prime swarm (see below).

What changes take place in the colony in the lead up to swarming?

As already noted, the queen is supposed to be fed less food and given the runaround by workers who are subjecting her to repeated vibration signals so that she is slimmed down in readiness for flight. As a result, the queen is supposed to be almost out-of-lay in the last few days before swarming. However, inspection of colonies at the time of swarming reveals that this behaviour is extremely variable. I have yet to see a colony in which there are no eggs at the time of swarming and in others, the queen can be seen to have been laying hard until almost the last moment.

This is just an observation and I have no idea what causes this variation or its significance. It may be yet another manifestation of lack of unanimity within the colony about swarming. Perhaps there is a shortage of bees which are trying to keep the queen's laying under control?

Work to rule

Some activities in the colony gradually go on 'work to rule' (they slow down) and this definitely includes an almost complete cessation of comb building. If frames of foundation have been introduced to the brood nest and they have not been drawn in a matter of a few days (even though there is a good nectar flow), this is a sure sign that the colony has already or is about to set up queen cells.

Traditionally it is claimed that during the build-up to swarming foraging is reduced but I have never found this to be obvious. It is also said that during the days leading up to swarming, many workers gorge themselves with honey and stand around doing nothing in particular, apparently waiting for the signal to go.

The first of these behavioural changes is not supported by the data derived from continuous monitoring of hive weight. This clearly shows that, providing there is a nectar flow and suitable foraging weather, a steady increase in hive weight is maintained right up to the point of swarming (so another beekeeping myth bites the dust?). There is also no clear evidence of bees gorging themselves well in advance of swarming, although some wax-makers may do so to enable the secretion of wax scales with little delay at the new nest site.

THE START OF QUEEN CELLS TO THE ISSUE OF THE PRIME SWARM

However, it has been shown that during the swarm initiation period (about one hour – see below), all the bees in the colony gorge themselves with honey, regardless of whether they are going to join the swarm or stay home. Presumably the bees that remain in the colony simply unload the honey when swarming activity has ceased and then resume their normal duties.

Honey carried by the swarm

I have been unable to find any data of the weight of honey that is carried by a swarm but it is often assumed that it is quite a lot (beekeeper paranoia?). However, if you do the calculation, assuming that the average weight that can be carried by a worker bee is 25 mg (dedicated foragers are specialised and can carry up to 40 mg), it shows that it would require 40,000 bees to carry 1 kg of honey. This means your typical swarm can only remove not much more than 250 g (or ½ lb) – which may be less than a large colony consumes on a non-foraging day in summer.

So swarming does not involve the major loss of the honey crop as some beekeepers claim.

Timing of the swarm

Swarming normally occurs soon after the first queen cells are sealed. It can occur earlier, well before any cells are sealed, but it is not known why – some colonies just do it! Peak time for swarming is 12.00–15.00 hours. Swarming can be delayed by poor weather and the colony gets more 'anxious' as time passes and may decide to leave at the first glimpse of sunshine.

The process of swarm initiation

Preparation for swarming (the warm-up period) takes at least one hour. The colony becomes increasingly active; there is an

Scout bees investigating a pile of empty equipment

explosion of vibration signalling and some workers emit piping noises. Finally, certain workers make what are called 'buzz runs' in waves across the combs and this is the signal for the swarm to emerge. These are the bees that lead the swarm to its initial clustering site where scouting continues. Authorities tend to disagree on how much scouting is actually done before the swarm emerges but this may not be a rigid pattern of behaviour. If for any reason scouting is a prolonged process, overenthusiastic wax-makers may deposit some of their wax scales and even start a sheet of comb at the initial clustering location.

Which bees control this process (initiate the swarm)?

Again it is not known but it is obvious that whoever is in charge must be in possession of two essential pieces of information:

- the maturity of the queen cells
- the prevailing weather conditions – is this a good day to swarm?

If it is the same bees that hold both types of information, it implies that they must be flying bees – bees that have knowledge of conditions both inside and outside the hive.

Looking at the behaviour as a whole, it seems likely that there is a controlling group of bees that orchestrates the swarming process. The most convincing evidence for this comes from artificial swarming. If all flying bees are removed from a colony containing queen cells (by moving it to a new position), the impulse to swarm is completely lost.

This evidence suggests that there is a group of bees that acts as 'swarm organisers'. They are obviously flying bees but are only a minority group (perhaps numbering as few as 100–200?) because the majority of the dedicated foragers seem to play no part in the process. Most foragers remain with the parent colony after swarming (see the next paragraph). Also, incoming foragers that get mixed up with an emerging swarm seem to quickly realise their mistake and abandon the swarm to return home.

How does the population of the colony split when it swarms?

This is the single most important piece of information as far as practical beekeeping is concerned. All too often beekeeping books say that a prime swarm consists of 'the old queen and the flying bees' – or imply this when dealing with methods of artificial swarming. Of course, bees are capable of flight from a very early age but, in this context, the term 'flying bees' is taken to mean bees that have reached the age when their main task is foraging.

Scientific studies of swarming show that this is totally incorrect and that a prime swarm is dominated by young bees, with up to 70% being under 10 days old and that includes individuals that are less than three days old (barely out of the cradle!). Many of the bees that emerge with the swarm will never have been outside the hive before but all they have to do

SWARMING BIOLOGY AND CONTROL

A group of 'swarm organiser' bees initiates the swarm and the other bees from the colony, including those that have never flown before, simply follow them

is fly and follow the swarm initiators and the queen. It has also been noted that on some occasions young bees that attempt to join the swarm have been overconfident about their ability to fly and fall to the ground in front to the hive. Exactly how the colony apportions its resource of bees between the swarm and the parent colony is not known.

At the point of departure, individual bees have to decide whether to stay or go and this does not appear to be predetermined. As all of the bees in the colony ('stayers' and 'goers' alike) gorge themselves on honey, this is not an indication that a bee is going to join the swarm. Also, if the swarming process aborts for some reason (usually because the queen does not leave the hive), the swarm will return home and try again later, but not all the bees that participated the first time will join the swarm for the second performance (and vice versa).

THE START OF QUEEN CELLS TO THE ISSUE OF THE PRIME SWARM

Seventy per cent of bees in a swarm are less than 10 days old

Artificial swarming is covered in Part 2, but it is the continued and persistent urge to swarm (come hell or high water) of the 'swarm organiser' bees that makes this management procedure so problematic. If the swarm organisers are not permitted to fulfil their role it may only be death that finally extinguishes their urge to swarm.

It is difficult to see how this knowledge about the composition of a swarm (which has been available for over 70 years) has failed to be incorporated into the British beekeeping culture until very recently. This information is readily available (*The Biology of the Honey Bee*, by Mark Winston, p 186), with supporting references going back to work done in the 1940s by Colin Butler at Rothamsted).

Natural swarm vigour

As far as beekeeping is concerned, this explains why a natural swarm is so vigorous and hard-working compared with a (beekeeper-created) artificial swarm. The natural swarm has an age structure that is ideally suited (fit for purpose) to the task it has to perform whereas the artificial swarm is just a gang of older bees, some of whom have to reinvent themselves to carry out the duties normally performed by younger bees.

Swarming is clearly a very complex and variable process and at the current state of knowledge, we are at a loss to understand some of the details.

> **The composition of a swarm**
>
> The question is why has this misinformation been so widely believed? If true, it would mean that a swarm was primarily composed of bees that were in the last third of their life span.
>
> To increase their payload, dedicated foragers shed up to 40% of their body weight and this is mostly achieved by atrophy of their glands – a condition that is only partly reversible.
>
> The mission of a swarm is to find a new nest site, build some comb (perhaps from scratch), forage and get the queen laying – not to mention collecting sufficient stores to survive the winter. Even when a new nest has been established, it will take at least 21 days before the swarm gets its first new recruit, by which time a high proportion of the (flying) bees would be dead and gone.

1.7 CAST SWARMING

An overview of cast swarming

With cast swarming we are entering the world of the unknown because this aspect of colony behaviour has received very little attention. This may be because most studies of swarming have made use of observation hives which, because of their physical size, contain a small colony that does not normally cast swarm and would certainly have an insufficient number of bees to issue a second cast swarm. If such a small colony did both swarm and cast swarm, the parent colony might be reduced to a non-viable size, so there must be a behavioural mechanism in place to prevent this from happening.

Cast swarming is not just another (rather inconvenient) swarm designed to frustrate the beekeeper, and there is good evidence that it is an integral part of the swarming process and makes an important contribution to its success.

A cast swarm is different from a prime swarm in a number of ways – not just by being headed by a virgin queen – but how the colony organises this is poorly understood. The following account involves some speculation but hopefully it is consistent with what can be observed by the beekeeper. It is also the sort of mechanism that is likely to have evolved through natural selection.

The advantages and (potential) disadvantages of polyandry were discussed earlier in Chapter 1.2. During cast swarming, the relationship between the queen and the bees that choose to accompany her in the swarm changes. It becomes more complex and, because of kinship recognition, opens the door to the exercise of **nepotism** – favouritism based on genetic affinities.

The honey bee colony seems to have strategies in place for avoiding or minimising the potentially disruptive behaviour that can arise through nepotism, but this is poorly understood at the present time. In terms of practical beekeeping, nepotism is of marginal significance. However, it can be a factor in relation to the handling of cast swarms and certainly helps explain their sometimes puzzling behaviour. There are also a few practical measures that can be used to increase the chance of some cast swarms developing into viable colonies (if that is what the beekeeper wants) and these are discussed in Chapter 1.8.

Although it is not immediately obvious, there is evidence that cast swarming is adaptive and contributes to the overall success of reproduction and the main explanation for this appears below. However, there are other aspects that have been deferred to Chapter 1.8 where they can be more usefully seen in the context of how the parent colony chooses its new queen.

The sequence of cast swarming

The following account of cast swarming describes what occurs naturally. It assumes that the beekeeper has not intervened in any way by reducing the remaining queen cells down to one. In other words, that after the prime swarm has departed there are multiple queen cells remaining and this situation has to be resolved by the colony, which it will normally do by cast swarming.

By building multiple queen cells, the colony has created a situation in which redundant queens have to be disposed of, one way or another. The choices are: being killed by a sister queen, killed by the workers (or by a collaboration between the two) or striking out on their own by heading a cast swarm. The final possibility, and what is the ultimate success for just one of these queens, is becoming the new queen heading the parent colony.

The virgins emerge

After the prime swarm has departed with the old queen, an average of about eight days elapses before the first virgin queen is ready to emerge – it can be more or less depending on the timing of the prime swarm. It will probably take 2–3 days before the chitin on a newly emerged queen's thorax has hardened enough for her to fly but when it has, and the weather is suitable, the first cast swarm will issue from the hive.

Prior to this, several virgin queens may have emerged (or been allowed to emerge), but others seem to be deliberately held back in their cells by the workers. It is generally assumed that the queens that have actually emerged will fight until there is just one survivor and soon thereafter the first cast swarm will issue. However, I doubt queen fighting is as simple as a free-for-all with the last queen standing winning because there is obviously some sort of underlying plan and the workers are at least in partial control.

Further discussion of queen fighting appears in Chapter 1.8.

Cast swarms with multiple virgin queens

If the colony is fairly large, a day or so after the first cast swarm has departed and after more virgin queens have emerged (or been allowed to emerge), a second cast swarm may issue and this not uncommonly contains more than one virgin queen (up to 14 have been recorded).

Despite the advice given in some beekeeping books, that when taking a swarm the queen should be found and safely held in a queen cage or matchbox, I have never even caught a glimpse of the queen in a prime swarm. However, in multi-queen cast swarms, queens can often be seen wandering around, mingling with the rest of the bees in the swarm. Because of their virgin status, they are subject to little or no attention from the surrounding workers and are probably keeping on the move to avoid something nasty happening to them. Such swarms are less tightly clustered and usually rather 'restless' with the bees moving around a lot – presumably because there is no mature queen on which to focus their attention.

Cast swarming in context

In nature (without the help of a beekeeper), a prime swarm has no better than a 20–30% chance of survival, ie, making it through its first winter. The survival of cast swarms has not, to my knowledge, been studied but with their inherent disadvantages – they issue about 10 days later in the season, contain fewer bees and have a virgin queen which has to mate before she can start laying – their chances of survival must be minimal.

On the positive side, the colony that produced the prime swarm (the parent colony) has a 90+% chance of producing a new laying queen and surviving until the following season. From this it is clear where the reproductive priority must lie and the whole process must have evolved to ensure that the parent colony is left in a condition which gives it the best chance of success (to be discussed further in Chapter 1.8).

Why do colonies cast swarm?

Surely it would be more adaptive to invest more (or all) resources into the part of the colony that has the highest probability of success?

Some beekeepers hold a very negative view of cast swarming, claiming that a colony can 'swarm itself to death'. However, I think this is just a simplistic explanation for colonies that fail to re-queen – something that can happen regardless of colony size. I suspect it is also used as an excuse for failing to notice that a colony has not re-queened after swarming in time to do something about it (it gives the beekeeper something other than his/herself to blame). Others claim that cast swarming is an artefact of beekeeping and only occurs with artificially large colonies. However, in my experience, feral colonies almost invariably issue at least one cast swarm, even from a nest site that is very small compared to a beekeeper's hive.

Another explanation for cast swarming that has been advanced is that it is a hangover from evolution in a tropical climate (where it has a much higher probability of success). However, there have been so many generations of cool-temperate-evolved honey bees that this does not sound like a plausible explanation.

Cast swarming must be adaptive to have survived for so long because, if it was any way non-adaptive, it would have been eliminated by natural selection long ago. So we need to take a positive view and see how it might benefit a honey bee colony.

What is adaptive about cast swarming?

The probable explanation is as follows:

- a swarming colony, almost by definition, has abundant assets in terms of bees, brood and stores – because this is one of the main triggers for swarming
- after the prime swarm has departed, the parent colony

still has a substantial reserve of largely sealed brood – so numbers are on the increase again
- having issued a prime swarm, the only aim of the parent colony is to survive so that it can do a repeat performance next year.

Viewed in this light, cast swarming can be interpreted as a means of temporarily reducing the number of mouths to be fed because this will help to conserve existing stores.

Also, because of the way the colony splits during swarming, the parent colony still retains the majority of the foragers which will continue to fulfil that role. The changes they have undergone to become foragers means that they are not well suited to perform other duties and they will continue to forage until they eventually die. During this period, until the new queen comes into lay, there will also be no brood to feed, so priority can be given to nectar collection.

The immediate needs of the colony post-swarming are to have enough worker bees to support the new queen, get her mated and tend her first batch of brood. The provision of these bees is the result of advanced planning, ie, increased laying by the queen in the period before the queen cells were started (see Chapter 1.6).

The next target for the colony is to rebuild its numbers, top-up its stores (if possible) and produce plenty of well-nourished winter bees.

There may also be another more subtle reason behind cast swarming, and that is the maintenance of social cohesion (discussed in Chapter 1.8).

Who organises the cast swarm?

The first question is, have all the original swarm-organiser bees left the colony in the prime swarm or have some stayed behind to run a follow-up event?

The indication is that all the original swarm organisers do, indeed, depart with the prime swarm. The evidence for this is that when a colony has been artificially swarmed and the parent colony (the part with the brood and the queen cells in it) is moved to a new position, all the flying bees will return to the artificial swarm (still in its original position). When this has had time to happen (probably no more than 24 hours in good weather), this part of the colony entirely loses the impulse to swarm.

The flying bees that relocated to the artificial swarm consist mainly of dedicated foragers but in their midst (and unidentifiable as far as the beekeeper is concerned) will be the swarm-organiser bees. Once these bees have departed, the parent colony will have completely lost its impulse to swarm and, no matter how many queen cells are present, it will settle down to establish a new queen (**and on no account will it swarm**).

New organising bees?

This implies that it is during the actual process of swarming – the period during which the bees are making the decision whether to stay or go – that a fresh group of bees is triggered to organise the cast swarm. So it appears to be a case of 'no prime swarm, no cast swarm', but how this works is not known. Could the departing swarm organisers do the equivalent of handing on the 'baton' or is it part of a hard-wired programme that automatically causes a group of bees to assess their remaining resources and decide whether further loss of bees would be advisable (or advantageous)? It is unlikely that this is an immediate decision because it is not the condition of the colony immediately after the prime swarm has departed that matters but rather how it is about 10 days later.

Like the prime swarm organisers, this new group of bees is very persistent and the accepted way to prevent a cast swarm is to give them no option by reducing the un-emerged queen cells down to one.

To be completely safe, this has to be done when there are no longer any eggs or young larvae from which the bees can construct additional emergency queen cells – which is something they do with great alacrity. The alternative is to reduce the queen cells now and come back later to check that no additional ones have been made and, if they have, destroy them.

If this precaution is not observed, the colony will happily cast swarm with the virgin queen that emerges from the single cell left by the beekeeper. It is only able to do this if the colony is aware that there are reserve queens in the emergency cells. No cast swarm will be produced until the colony has assured itself that there is at least one un-emerged virgin queen left in the colony and queen piping is almost certainly one way of doing this (see Chapter 1.8).

Queen mating

When the process of swarming is over, all cast swarms have departed and the colony has selected its intended queen, it is then simply a matter of getting her mated and into lay as soon as possible. This has already been discussed in some detail in Chapter 1.2 and will be further discussed in Chapter 1.8.

1.8 SELECTION OF THE NEW QUEEN AND KIN RELATIONSHIPS

The normal colony

In a normal (mature) colony where the queen has been in residence for several weeks, all the worker bees are her daughters. They therefore have an equal 50% genetic stake (the 50% of their genes that came from their common mother) in her success so there is no basis for competition or any conflict of interest. The same is true with the prime swarm; all the bees in the swarm are equally related to the queen and have a common interest in her survival.

As already noted in Chapter 1.2, the workers in a colony belong to a number of sisterships (patrilines), thus creating a complex system of family relationships. There is evidence that nurse bees have a slight tendency to feed larvae who are their full sisters in preference to those who are half-sisters – but that is all. In practical terms, this low-level nepotism (favouring members of your own family) is not significant and all the larvae get fed properly.

However, when it comes to swarming there are really big genetic advantages to be gained if a given patriline can ensure that the queen which inherits the parent colony is a full sister with whom they share 75% of their genes (25% from their mother and 50% from their drone father). However, there is no evidence that some patrilines do not join the prime swarm because they know that some of the queen cells contain their full sisters and stay behind to promote them (that possibility has been ruled out by previous investigation).

What is the evidence for nepotism?

When bees set up to swarm there is evidence in some colonies (but not others) that certain patrilines appear to be numerically over-represented in the queen cells that are started – again this could be a form of nepotism. But why in some colonies and not others? This may simply be due to the fact that some patrilines are more numerous than others in the fertilised eggs that are being laid at that particular time.

A certain amount of patriline inequality inevitably arises at the time when the queen is stocking her spermatheca, when there is no means by which she can ensure that she stores equal contributions from every drone.

Queen failure is often blamed on poor mating and one cause is claimed to be infertile drones; drones that produce sub-standard or non-viable sperm. A newly mated queen temporarily stores all the semen that she has received in her median oviduct. During her maturation period (before she starts to lay), she gradually pumps this out and 90–95% goes to waste and is cleaned-up by attendant workers. As the sperm passes the entrance to her spermatheca, it has to actively swim into the spermathecal duct in order to be stored.

For this reason, drones producing non-viable sperm seems to be an unlikely explanation for queen failure because their sperm would never reach the spermatheca. Of course, some sperm may be viable at the time it is stored but become non-viable during storage. At least initially, that is only likely to produce pepper-pot brood.

The accepted scientific view is that the queen homogenises her store of sperm, in which case paternity should have a random distribution. Appearing to partly contradict this is the fact that some studies have shown that sperm mixing increases with time – but without saying how long it takes for mixing to be complete. Other genetic studies of colonies show that the profile of the patrilines (in terms of their numerical representation) often changes with time.

Further anecdotal evidence comes from beekeepers who claim that, whilst still headed by the same queen, they have seen quite marked changes in colony colour balance over time.

Taken together, these observations suggest there is some clumping of sperm in the queen's spermatheca.

It is after the prime swarm has departed that potential conflicts of interest arise. 'Potential' is really the wrong word because such conflicts are a biological reality and **must** exist. But how the colony overcomes the situation without fighting between patrilines (equivalent to civil war), is where the interest lies. At the present time there are no clear answers.

After the prime swarm

If the prime swarm has left on time as the first queen cells have been sealed, there are about eight days to run before the first virgin queen emerges. Most books say that she immediately starts to pipe and that mature queens still in their cells pipe back – but the sound is muffled by the cell walls so it is usually referred to as 'quacking'. This is said to enable the emerged queen to locate her rivals, chew through the wall of each queen cell and sting the occupant to death (getting rid of the competition).

This is clearly an over-simplification and what you usually see in a colony at this point in the process is several emerged queen cells and a lot more that remain intact. These un-emerged queen cells are presumably being protected by workers who must have a positive reason why they want them to survive. It is not because the queens they contain are not yet mature because many of the cells have a line round the tip where a mature queen is actively trying to chew her way out but is being prevented from so doing by the worker bees ('warder bees'). During this period of 'incarceration' the queens-in-waiting are fed by their attendant workers.

As already noted in Chapter 1.7, two to three days after the first virgin emerges, the colony will issue a cast swarm but that will only occur after a few emerged queens have fought and there has been a single victor. The successful queen may have destroyed a few unprotected rivals still in their cells during this time but, when she is able to fly, she will issue with a swarm. Most first cast swarms seem to contain only one virgin queen.

If the colony is quite large, a few days later it will issue a second cast swarm and experience shows that this often contains two or more virgin queens – up to 14 have been reported. This suggests that with later cast swarms (or possibly it is only the last cast swarm?), the queens do not (or are not permitted) to fight within the colony to the stage where there is a 'last queen standing'.

So now you have a swarm in which the queen that will eventually be dominant has not been resolved in advance. This implies that later cast swarming operates in a different way, but one can only speculate as to why this is so.

The difference between cast swarms

One possible cause for the difference between the first cast

SELECTION OF THE NEW QUEEN AND KIN RELATIONSHIPS

What is the role of queen piping?

Many people seem to regard queen piping as a sort of challenge, possibly because the sound is vaguely militaristic or simply because it can result in conflict of a sort. However, in nature most challenges are a means of avoiding conflict not promoting it. The rivals confront one another and perform what is usually a highly ritualised challenge and by this means assess their likelihood of winning. Usually one challenger backs down and lives to fight another day. Only rarely is there full-scale conflict.

So why, if you are a queen imprisoned and defenceless in a cell, would you want to reply to a challenge from an emerged queen? This would be nothing other than signing your own death warrant. My view is that piping by un-emerged queens is targeting the workers, not the emerged queen. Workers stop in their tracks when piping occurs. Perhaps this is to feel the vibrations better and possibly make a fitness assessment – a sort of piping competition between un-emerged queens?

I see this as a means by which un-emerged queens can display their fitness to the workers. The sound is made using the wing muscles and a good queen needs strong wing muscles. So what do they want from the workers? The most likely thing is protection from this marauding virgin queen which is hell bent on killing them. Maybe it is also part of a longer-term strategy to be chosen as the queen which inherits her mother's colony – see further discussion below. It has been suggested that a colony uses the piping of queens still in their cells to detect/confirm their presence and thus allow an emerged queen (or queens) to issue with a cast swarm knowing that a backup is available. However, this cannot be the only mechanism because a colony in which occupants of backup queen cells are too immature to pipe can also allow a cast swarm to issue.

swarm and later ones is the maturity of the virgin queens at the time of emergence. The first virgin queens to emerge will be just full-term (around 16 days old) and until their chitin has fully hardened they will be unable to fly, thus enforcing a delay of 1–3 days before cast swarming is possible.

Presumably queens which are prevented from emerging by the workers continue to mature in their cells and may be capable of flight almost immediately after emergence. This may reduce the time available for queen fighting and/or make it easier for the workers to keep rival queens apart – if this is what they want to do.

So what is going on here? There must be competition between the virgin queens and the workers are now in a situation where it is possible for them to be nepotistic, ie, favour a queen to which they are a full sister or are more closely genetically related. This is a potentially divisive situation and the repercussions are clearly evident in the behaviour of some multi-queen cast swarms.

Behaviour of cast swarms with multiple queens

Experience of dealing with cast swarms shows that this is very variable and the following is the list of possibilities (all of which have been seen to actually occur):

- The bees can split up as the swarm issues from the hive and sub-swarms fly off in different directions.
- The bees can settle nearby and form a cluster, or sometimes more than one cluster. If it is a single cluster there are usually signs that the bees are not completely 'comfortable', and the cluster is not as tight as normal (the workers are moving around a lot). Shortly thereafter the cluster can split and part of it fly away.
- If the beekeeper has gathered the swarm in a box or skep, part of it can abscond after a short while and before there is time to put the whole swarm into a hive.

SELECTION OF THE NEW QUEEN AND KIN RELATIONSHIPS

A cast swarm containing multiple virgin queens settles in two groups, with another swarm inside the box

- When introduced to a hive, parts of the swarm can occupy different frames as far away from each other as possible. If left in this condition, one part can abscond at any time over the next couple of days.
- Finally, they can apparently reach some sort of agreement (consensus) and combine. This can often be 'encouraged' if the beekeeper shakes all the bees off the frames into a heap on the floor of the hive thus forcing them to come to a mutual agreement. Even this does not always work and the bees will re-segregate based on their favoured queen and one sub-swarm will depart. The only thing I have never seen is two sub-swarms occupying different parts of the hive, long-term.

There is really only one explanation for this behaviour and that is nepotism. The bees in the cast swarm obviously favour different queens and are prepared to go to considerable lengths to stick to their choice. Although this has not been proven scientifically, the basis of this must be the result of kinship recognition, with bees that are closely related to the queen of their choice wanting to ensure her success by establishing a new colony.

In reality they are 'shooting themselves in the foot' because the already slim chance of a cast swarm surviving has been further eroded by creating still smaller sub-swarms. So, as far as they are concerned this behaviour is non-adaptive, so why does it persist?

Seen in the wider context of successful reproduction, the non-adaptive behaviour of multi-queen cast swarms can still be regarded as adaptive for the colony as a whole. So, although nepotism can be seen to be alive and well under some circumstances, its expression has been marginalised and is not permitted to undermine social cohesion and harmony in the all-important parent colony (see below).

Choice of a new queen by the parent colony

After cast swarming is complete comes the choice of the new queen for the parent colony and to achieve its aim of surviving and passing on its genes, there are some harsh realities.

It has already been noted that in nature (without the help of beekeepers), a prime swarm has no better than a 20–30% chance of survival, ie, making it through its first winter. At first sight this may sound like a poor outcome but this level of survival is probably enough to ensure population stability. If the success rate was much higher year-on-year, we would become overwhelmed by honey bee colonies! However, the colony population in a given area is probably limited by other factors such as the availability of good quality (survivable) nest sites. If there is a large number of nest sites in an area, then the availability of forage might become the next limiting factor.

It has already been noted that the survival of cast swarms has never been studied but, given the fact that they are smaller,

Undersized queens

In the case of some small cast swarms, which are presumably secondary in nature, ie, not the first cast swarm, it is not uncommon to find that they contain an undersized queen. She may even be an inter-caste, ie, a queen which, due to lack of adequate feeding during her development, is undersized and does not have the full complement of ovarioles (often referred to as a 'scrub' queen). Although they sometimes mate successfully, such queens never succeed in building a fully functioning colony and often die prematurely.

A more definitive example of this was seen when one of our colonies issued a cast swarm which atypically settled on the ground about 10 m from the hive. It was collected and hived in a nucleus box.

It was only a few days later, when the queen went out to mate, that we became aware of the problem. She was found on the ground beneath a small cluster of workers and, on examination, was seen to have normal wings on one side of her body but those on the other side were half-size. Not surprisingly, she was only able to fly in circles.

The parent colony from which this swarm issued still had plenty of good-quality un-emerged queens available, so the question is, why did these workers decide to swarm with a defective (no-hoper) queen? Why had she not been eliminated by other fully developed queens – or by the workers? What was the motivation of the workers that were prepared to swarm with her? Was this an extreme (and ill-advised) case of nepotism? This just underlines how complicated some aspects of swarming can be and how unwise it is of us beekeepers to interfere unless it becomes absolutely necessary.

issue later in the season (10–12 days) and have a virgin queen which has to get mated before she can start to lay (another 10–14 days perhaps), a low survival rate is not surprising. This means that a cast swarm has a total of 20–26 days (a whole worker brood cycle) disadvantage compared with a prime swarm. So the chance of survival for a cast swarm in a temperate climate with a fairly short season must be seen as minimal.

It has already been argued that cast swarming has advantages for the parent colony's economy – fewer mouths to feed at a time when it is not essential to have a large number of bees. So, although it is unlikely to result in a new independent colony, cast swarming can be seen to play an important role in the overall success of swarming. There may also be another hidden advantage of cast swarming because it could act as a sort of genetic safety valve for the parent colony by exiling malcontents – a means of preventing conflict between the patrilines and thus ensuring colony cohesion (see below).

On the other hand, the virgin queen which inherits the parental home has by far the best chance of long-term survival. She has a ready-made home and plenty of other resources in terms of bees, the remains of her mother's brood, and stores, and all she has to do is mate and start laying. This queen has a 90+% chance of survival and to still be heading the colony a year hence.

From this it is clear that this is where the reproductive priority must lie in the swarming process as whole – **it should ensure that the parent colony has the optimum conditions for success**.

So how is competition between queens in the parent colony resolved?

Competition is not just some airy-fairy theory, this is how it is in the real world – 'nature red in tooth and claw', as they say. But with a bee colony, apart from a few fights between

SELECTION OF THE NEW QUEEN AND KIN RELATIONSHIPS

queens, the whole issue seems to be resolved peacefully and, to outward appearances anyway, the colony continues to function normally throughout the swarming process.

Unlike emergency re-queening, where the workers themselves typically tear down 50% (or more) of the queen cells before the first virgin is ready to emerge, in the case of swarming the workers permit a majority of the queen cells to develop to maturity.

Choosing the queen

The most likely strategy by which the colony chooses the all-important new queen (that will ensure its survival) is that it is not left to chance by simply allowing the remaining queens to fight for supremacy. It seems likely that the colony has a favourite queen (or more likely a shortlist of queens) which is held back in her cell and protected from marauding, emerged virgins, and continued to be fed until cast swarming is over.

It is not clear what happens next but, if the colony has held back multiple queens, then they are probably allowed to emerge and fight until there is just one queen remaining. There is observational evidence that the workers do not just stand idly by during queen fighting but often play some part in determining the winner – this is where the phenomenon of queen balling is involved. Worker bees have been observed to hold one queen down while her opponent stings her to death – in other words a fixed fight.

There are also some poorly understood interactions between the combatants themselves, where one queen can break off the fight by spraying her opponent with a liquid of unknown composition, ejected from her anus. This is thought to be a one-off event and the fight may be resumed later, but how this affects the outcome and where it fits into the process of selecting a new queen is not understood.

The question remains, how do the bees choose the queens that remain after cast swarming? Is it based on some measure of their fitness? Or is it the result of nepotism, eg, does a dominant patriline contrive to promote a queen which is their full sister? Maybe it is the workers' assessment of the queens' piping performance (see discussion above)? Investigations using observation hives make it clear that the workers are losing interest in some queen cells whilst others are attracting most of the attention, but they fall short of being able to identify the eventual winner in advance. The answer is that we do not really know.

Behaviour in the presence of queen cells

Unfortunately, the literature on the way that bees behave in the presence of queen cells is both confused and confusing.

Some of the experiments to investigate this have been done using emergency queen cells that have been produced after removal of the queen. This is because it is not easy to induce colonies to produce genuine swarm cells to order and experimenters want to do things according to their time schedule (not the colony's) – and usually with several replicates. Also, in the attempt to control variables that are not involved in the investigation, research sometimes strays outside the parameters of normal colony behaviour and the experiment no longer represents a real-life situation.

It is obvious that some researchers in this field do not fully appreciate that emergency re-queening and swarming are controlled by two entirely different behavioural programmes with different aims and outcomes and different methods of achieving them. The most obvious difference is that when a colony is undergoing re-queening by the emergency programme, it has absolutely no intention of swarming and will never do so no matter how many queen cells have been produced.

In a more extreme example of poorly conceived research, the colonies being studied had their queens removed and were then given ready made queen cells that had been produced by grafting using a different colony. What behavioural programme these colonies were following is anybody's guess but it certainly had little to do with swarming!

The bottom line is that there is a great deal of interaction between worker bees and the maturing queen cells which involves both vibration signals and pheromones. The bees presumably know what it's all about but we don't. As is always the case with honey bees, we mere mortals should assume that, through natural selection, they have evolved to do it in the best possible way (for them). For the beekeeper, it is wise not to interfere and to leave the bees alone to get on with it.

It would therefore seem that the traditional advice that hives should not be opened during the period from when queens should have started to emerge up until there are signs that the new queen has started to lay, is very sound. This covers the time when the new queen is being chosen and when she is likely to be making mating flights. During the later part of this period, when only mating is in progress, a hive can be carefully opened outside of mating hours (before 10.00 hrs and after 16.00 hrs) but for essential purposes only (like adding extra supers) and it is better not to venture beneath the queen excluder.

Management of cast swarming

Some of the complexities of cast swarming have been discussed above (which is all very interesting) but, from a beekeeping standpoint (ie, to preserve your precious honey crop), it is important to prevent cast swarming.

The conventional method by which this is achieved is that, after the prime swarm has departed, the remaining queen cells are reduced down to one. If there is still worker brood of an age from which emergency queen cells can be started, the colony must be checked a few days later and any new cells destroyed.

Reducing the number of queen cells is seen as a very undesirable practice because it means that the beekeeper (not the colony) is choosing the new queen. Of course, this happens all the time in conventional practice when the beekeeper chooses to replace a queen and usually introduces a queen from a different genetic source. This is something else again and totally unnatural and is, in effect, killing the colony.

Releasing virgin queens

An alternative method for preventing cast swarming (which we now use all the time) is to estimate when the queen cells will be mature and queens will start to emerge. To achieve this it may be necessary to open a few cells to assess the maturity of the occupants. We then return on the calculated date and, if we have got it right, we will usually find that one or two queens have already emerged. Then (using a scalpel) we release as many virgin queens into the colony as we can (or until we get bored). When that has been done, we destroy all remaining queen cells so there is no backup queen available.

SELECTION OF THE NEW QUEEN AND KIN RELATIONSHIPS

Releasing a virgin queen into the colony

In the past, we have released 18 virgins into a colony that must have already had at least one present. We have no idea what happens next (or how it is accomplished) but consistently we find that the colony will choose one queen from those running loose in the hive and make no attempt to swarm. This is done without any apparent conflict or loss of blood – except of course that all but one of the queens is eliminated by some means.

This rather counter-intuitive form of management seems to tap into an instinctive colony behaviour, which is presumably the one used when the colony decides it has issued enough cast swarms and must settle down and chose a new queen. Having no un-emerged backup queen is probably the key to the success of this manipulation. The choice of queen may not be identical to that which would have occurred naturally but it is better than the beekeeper making a unilateral decision on behalf of the colony.

Should the beekeeper interfere?

One final question remains. It has been argued above that cast swarming is adaptive for the colony as a whole, so is it wise to interfere by preventing it from happening? I have never seen any evidence of that in a beekeeping context. If the main advantage to the colony is resource economy (reducing the number of mouths to be fed), then the beekeeper can make good any shortage during pre-winter feeding.

It is ironic that practising good swarm control and successfully maintaining a large colony for the entire season can sometimes work against the beekeeper.

In years when there is a good spring flow but the later one is a failure (poor weather in July), your diligent beekeeper can end the season with a meagre crop of honey. At the beginning of June the hives will have contained a substantial crop of honey but by the end of July a large colony will have consumed most of it. By contrast, a beekeeper who does not care a fig about swarm control and has let his/her colonies both swarm and cast swarm, so that they have few mouths to feed for the latter part of the season, enjoys a reasonable crop. Under these circumstances, cast swarming can also be seen as adaptive for the beekeeper!

Conclusions

Well, there really are no conclusions except that the choice of the new queen for the colony is subject to complex behaviour, so it is just as well the bees know what they are doing because we as sure as hell don't. So the best policy is, as far as possible, to let them do it their way.

I should emphasise that this should not be taken as advice not to practice swarm control but merely to let the colony choose a new queen in its own way and from the largest pool of candidates possible.

PART 2

PRACTICAL SWARM CONTROL

2.1 PRACTICAL SWARM CONTROL

In their book entitled *Bait Hives for Honey Bees*, Thomas Seeley and Roger Morse state that, 'Mature colonies have a natural urge to swarm each year unless weakened by disease or mismanagement'. So perhaps we should not be surprised or regard it as dysfunctional when colonies swarm. Most beekeeping books understate swarming but its control is vital if good honey crops are to be consistently obtained. There is no doubt that swarm control is simultaneously the most important and most difficult aspect of colony management.

Hopefully, the earlier chapters of this book, which cover what is known about the biology of swarming and also advance some working hypotheses for what is not known, should help the practical beekeeper to better understand the management involved. Many beekeeping manuals are overprescriptive and I think it is far better to understand the underlying principles which will enable you adapt to the many and varied situations that beekeeping presents. However, anybody who claims to have complete mastery of swarm control is either a poor observer, a liar or not of this world.

In the highly unpredictable UK climate, some years are quite easy and only a few colonies set up to swarm, whilst in others virtually every colony will make an attempt at some time during the season.

Types of swarm control

Many texts about swarming use the terms 'control' and 'prevention' almost indiscriminately without any definition and this clearly results in confusion.

From the point of view of practical management, swarm control can be divided into two distinct parts with a clear (biological) threshold between them **which is when the colony starts queen cells**. What follows is a clearer nomenclature for the two types of swarm control:

1. *pre-emptive swarm control* – the management techniques that can be used before queen cells are present in the hive (to try to prevent their initiation)
2. *re-active swarm control* – the management techniques that can be used when queen cells are produced (to prevent the issue of swarms).

Some beekeepers seem to have misunderstood the clear distinction between these two types of management practices and apply pre-emptive management after queen cells have been started in the hope that it will make them go away. It is a vain hope I am afraid and there is very little chance this will cause the colony to change direction. The Demaree method of swarm control is often the subject of misunderstanding because, as described by its originator (*American Bee Journal*, 1892), it is clearly pre-emptive. However, a range of modifications have been proposed over the years which has created much confusion.

Queen clipping

Clipping a queen's wings on one side of her body so that she

PRACTICAL SWARM CONTROL

is unable to fly (at least not far or in a straight line) is not a method of swarm control. It is merely a means by which a beekeeper, wishing to practice swarm control, can reduce the required frequency of hive inspections (looking for the start of queen cells) from about 7 up to 14 days without the risk of losing a prime swarm.

When a colony with a clipped queen attempts to swarm, the queen falls to the ground (usually in front of the hive). The swarm will initially cluster around her but after a while abandon her in disgust and return to the hive.

A swarm of bees clustering round a clipped queen that is unable to fly and has fallen to the ground

This does not mean they have lost the urge to swarm because they will simply wait for the first virgin queen to emerge and be ready to fly and then repeat the swarming process – often involving the loss of more bees than with the aborted first attempt. From the start of queen cells to the earliest possibility of this second swarm is about 18 days, so a 14-day inspection period has a good safety margin.

SWARMING BIOLOGY AND CONTROL

Most beekeeping books recommend queen clipping as standard practice but without any attempt to look at the pros and cons for either the beekeeper or the colony. I also realise that some people have strong ethical views about mutilating queens, which I respect.

For the colony, the practice of wing clipping incurs serious disadvantages. If the colony gets as far as issuing the prime swarm (and that can easily happen in seven days or less) the old queen is usually lost and that means termination of one branch of its genetic line. If the parent colony then fails to re-queen, the genetic line becomes extinct and this is the ultimate failure as far as the colony is concerned.

Clipping the queen's wings on one side means she is unable to fly with the swarm

Queen clipping also has several serious downsides for the beekeeper which never seem to be discussed by those who recommend the practice. Unless the apiary is patrolled daily during the swarming season in order to recover the non-flying queen from the ground, a valuable mated queen has been lost along with her laying potential. The colony will not have a new laying queen for at least three weeks and the beekeeper still has to deal with the problem of cast swarms. This creates a substantial brood break and, after the existing brood (produced by the old queen before her departure) has emerged, it will be at least two weeks and possibly four weeks before any new recruits are produced. During this period, the number of bees will inevitably decline and with it the potential honey production.

Routine inspection for swarm control

There is no way of escaping the need for regular colony inspections if the beekeeper has decided to practice rigorous swarm control and, unless you have decided to practice wing-clipping of queens (see above), there is no safe option other than to inspect colonies every seven days (or thereabouts).

Even this is not completely safe because colonies can (and sometimes do) go from no sign of queen cells to swarming in as little as four days with not a sealed queen cell in sight. Sometimes this can be induced by the beekeeper destroying queen cells in the hope of preventing or delaying swarming but it also happens completely out of the blue. There is nothing sensible that can be done about this – it is just one of the unpredictable situations that occur in beekeeping.

The swarming season generally extends from sometime in the middle of March to the middle of July but this does vary in different parts of the country and particularly in relation to elevation. It also varies from one season to the next (there are early seasons and late seasons) and it is determined by the developmental state of the colonies.

In order to plan your inspection routine, you have to keep an eye on colony development, starting in March so that you know when to start routinely checking for queen cells. You should also be aware that strains of bee with a Mediterranean origin have a second (but usually less marked) swarming season in the late summer (August and September). The only swarms we have seen in that period have all been more yellow than our locally adapted, near-native bees.

Having decided on the likely swarming season for the current year, colonies need to be inspected at roughly seven-day intervals – or as near to that as other commitments and the weather allow.

Opening the hive

All sorts of arguments are used to suggest that each time a hive is opened it is a traumatic experience for the colony, from which it may take days to recover. The clinching argument (deterrent) for many beekeepers is that it is often claimed that every time you open a hive it reduces the honey crop.

The idea that opening hives has a prolonged effect on the hive atmosphere from which it takes time to recover can be seen to be completely false. Firstly, the chemical signals that are used to control activity in the hive (including the complex mix of pheromones produced by the queen) are all large molecules with low volatility and they deliver their messages through physical contact between the bees or from the surface of the combs. The information contained in the messages is intended to have a fairly immediate effect and be capable of being switched both on and off. Therefore it has to have a short active life and needs constant renewal at source.

This sort of communication simply would not work if the message remained persistent in the hive atmosphere. It should also be realised that an active colony in summer is typically exchanging air at a rate of 40–60 litres per minute – and a rate of up to 200 litres per minute has been recorded. Compared with that, what difference does the relatively brief opening of a hive make?

We open our own hives on a more-or-less weekly basis during the swarming season to check for queen cells. All our hives are on a two brood-box configuration, so once the comb management has been completed for that season, inspection is limited to lifting one side of the upper brood box to see if there are any queen cells on the bottom bars. If anything suspicious is seen, a full inspection is made. This is not absolutely 100% reliable but is safe enough and saves a lot of time if you have a number of colonies to check. It is very rare to find queen cells in the bottom box and none on the bottom bars of the frames in the upper box.

A colony intent on swarming

My conclusion, based on years of managing colonies fairly intensively, is that opening hives affects them much less than is often suggested. However, there is one important proviso and that is that great care must be taken to ensure that all parts of the hive have correct bee-space and that it is genuinely a movable-frame hive. Incorrect bee-space is a nightmare for the beekeeper and even worse for the bees. They are also likely to communicate their displeasure about this situation to you – and who can blame them!

SWARMING BIOLOGY AND CONTROL

If during your inspection you have found queen cells in a colony, the first thing you have to do is find out exactly what is going on (is it swarming or something else?). If you decide it is swarming, what stage in the process has been reached? This is covered in the form of a diagnostic key in Chapter 2.3.

> **Weekly hive inspections**
>
> We currently run our Association's teaching apiary which has to be managed quite intensively to ensure that the maximum number of colonies are available for teaching purposes.
>
> For the beginners, the aim is that there should be no 'complications' that have to be explained too early in what is a progressive learning process. To meet these requirements, the hives have to be opened weekly and, on teaching weeks, they can be opened up to three times. At least half the inspections during the season are done by raw beginners (under supervision, of course) and a hive may be open for up to one hour (shock horror!).
>
> Despite this apparently harsh regime, the association colonies are healthy, produce a good crop of honey and appear to suffer no ill effects.

Swarm control and honey production

The aim of most beekeepers is to produce honey. This may be just enough for friends and family or, at the other extreme, as much honey as possible for sale, thus creating an income stream that offsets the cost of beekeeping and even makes a profit. In other countries they have categories such hobby or backyard beekeepers (with up to about 10 colonies), sideliners (with up to 100 colonies – to generate some income) and, finally, commercial beekeepers or bee farmers (with over 100 colonies – to be a main source of income).

The smaller the number of colonies the beekeeper runs the more close attention to swarm control is feasible. Clearly a bee farmer cannot inspect colonies every seven days during the swarming season to check whether swarm cells have been started and they have to rely on pre-emptive swarm control. But, if they have the time and inclination, hobby beekeepers are able to do regular inspections.

Alternative swarm control strategies

There are five main strategies in relation to swarm control that the beekeeper can adopt. Which is likely to be the most successful, ie, prevent the loss of swarms and produce a good crop of honey, varies from year to year depending mostly on the sequence of the weather. A crystal ball comes in handy here!

1. *Try and keep the colony together for the whole season.* If this is successful, it will result in a good honey crop but, if it fails and the colony sets up to swarm, the beekeeper has to change to strategy (3) or (4).

2. *Split the colony (a pre-emptive split) at a carefully chosen point in the season.* This will also result in a good honey crop which, under some circumstances, can be better than strategy (1). The 'carefully chosen point in the season' is based on doing the split before the colony tries to do it for itself, ie, sets up to swarm, and in relation to potential nectar flows. One obvious example is splitting the colonies in the late spring after autumn-sown oilseed rape has ceased to flower, thus allowing time for recovery before the main flow later in the season.

3. *Wait until queen cells are started and then do an artificial swarm.* This usually results in a reduced crop of honey but is much better than losing a swarm (or swarms).

4. *Allow the colony to swarm.* Catch and hive the swarm and prevent cast swarming. Because a natural swarm (as opposed to an artificial swarm) has an ideal age structure, it will make rapid progress and, if swarming

occurs early in the season, the swarm has the potential to produce a significant crop of honey. If the parent colony is prevented from cast swarming, it, too, can produce a significant crop and the combination can often be better than strategy (3). However, the difficult part of this strategy is ensuring that you catch the swarm! Of course, you could sit in the apiary every day from 11.00–16.00 hours during the swarming season to make sure! Or (more realistically) you could set up bait hives and inspect them regularly (see Appendix 5).

5   *Let-alone beekeeping (laissez-faire).* Do little or no management of the colony apart from adding supers and hope for the best. You may get lucky and the 'best' will happen – but not very often. However, if you are just interested in the bees, don't care about honey and live in a rural area where loss of swarms is not a public nuisance, this is a viable option. But it does beg the question, why are you reading this book?

Use of bait hives

If the beekeeper chooses (or only has the time) to adopt swarm control strategies 4 or 5 (allowing the colony to swarm or let-alone beekeeping), setting up bait hives in the hope of catching swarms solves a number of problems. Catching a swarm and getting it properly established in a hive may increase your honey crop, provide spare colonies (for you or for others) and can also be an ideal way to get a supply of new drawn combs. Appendix 5 contains guidelines for how best to do this, from what is most likely to attract a swarm right through to its management when you have been successful.

Triggers for swarming

As a reminder, the triggers for swarming – the means by which the colony recognises when it is a good time to swarm and is most likely to be successful – are multifactorial and a mixture of internal and external conditions. This has already been discussed in greater detail in Chapter 1.5 but it is useful to reiterate it here in a slightly different format which clearly separates internal and external factors.

Internal factors (within the hive)

- The size of the colony, the space for the queen to lay, brood nest congestion, brood nest maturity and (possibly) the age of the queen.
- Production and/or distribution of queen substance is thought to be the main mechanism.
- Space for nectar processing and honey storage.
- Hive ventilation – not really a problem with open-mesh floors (providing you leave the catch-tray out during the summer).

External factors

- Time in the season – the swarming urge is at a peak in May and June (when it is most likely to be successful) and declines thereafter.
- Weather – an underrated factor as interludes of poor weather (with little flying time) often precipitate swarming. Excessive temperatures are rarely a problem in the UK but good hive ventilation (see above) solves that one.

To some extent the beekeeper can control internal conditions through management of the hive but can do nothing about the external factors. It follows that pre-emptive swarm control is mostly about management of the brood area, making sure there is always space for the queen to lay. Providing there is adequate space for a colony to process nectar and store honey, intensive management of the colony above the queen excluder is less important than that below.

SWARMING BIOLOGY AND CONTROL

The main management techniques by which the beekeeper can control internal hive conditions are:

1. comb management
2. box management
3. brood relocation
4. splitting colonies.

Details coming up next, in Chapter 2.2.

2.2 PRE-EMPTIVE SWARM CONTROL

THAT DARNED BEEKEEPER SPOILED OUR PLANS AGAIN!

There are some wider implications connected with the practice of pre-emptive swarm control – I hesitate to call them downsides – of which I think you should be made aware.

Firstly, all but the pre-emptive splitting of colonies (see the end of Chapter 2.1) will have the effect of producing a larger colony – which is what the beekeeper wants for maximum honey production. However, the larger the colony the more likely it is to swarm. So pre-emptive swarm control can be seen as something of a Catch-22 situation (dictionary definition: *a dilemma or difficult circumstance from which there is no escape because of mutually conflicting or dependent conditions*). That sums it up nicely!

The second problem is that, although pre-emptive management always delays a colony attempting to swarm, it may fail to do this for the whole season.

SWARMING BIOLOGY AND CONTROL

It is not uncommon for it to fail sometime in late-June or early-July just when (in most areas) the main nectar flow is starting. At this time, the last thing the beekeeper wants to create is an artificial swarm, which splits the colony and reduces its potential as a foraging force. There are ways of dealing with this eventuality and these are discussed in Chapter 2.4.

Splitting a colony is the most powerful method of pre-emptive control and, if done with the correct allocation of resources, it usually prevents swarming for the rest of the season.

Accompanying diagrams

As well as text instructions for the various methods, each is accompanied by labelled diagrams. For the sake of simplicity, all the diagrams assume the use of a Modified National hive with 12 self-spacing (Hoffman) frames. However, the same management techniques can be applied to all types of movable frame hive whatever the size or shape and number of frames.

The frames shown in the diagrams are colour-coded according to their contents:

- frames containing brood are coloured red
- honey and pollen are yellow
- drawn frames with no contents are black
- dummy boards are blue
- un-drawn frames (containing a sheet of foundation) are shown as a thin black line.

In the real world, the frames within a hive usually have a mixture of contents so the colour coding refers to the primary contents (the dominant characteristic) of the frame.

Most diagrams also show a standard hive configuration of a brood-and-a-half, ie, one deep and one shallow box beneath the queen excluder (the brood area), which seems to be about the right amount of space for near-native (locally adapted) bees here in Wales. Other hive configurations (eg, single deep, double deep or 14x12) require appropriate modification of the methods described below but some options may be limited or even impossible.

2.2.1 Comb management

The aim here is to ensure that as many as possible of the frames below the queen excluder (the brood frames) are actually used for the production of brood. During the main season, stores of honey and pollen in the brood area should be kept to a minimum.

Quite early on in the season, when the colony has not yet attained its full potential size, it is the bees' instinct to create a ceiling of capped honey. These (close-to-hand) stores are a form of insurance against adverse conditions and the bees are reluctant to uncap them and make cells available for the queen to lay. When this ceiling is in place, the only way the brood nest can be extended is in a downward direction, so the aim should be to have brood in contact with the queen excluder over as much of its area as possible. The bees will still create a honey ceiling but it will be in the first super – which is where the beekeeper wants it to be.

Comb management may also involve moving existing frames within the brood area in order to provide space for the queen to lay but the main activity is removing old or defective frames and getting new frames of foundation drawn. It is good beekeeping practice to replace brood frames on roughly a three-year cycle, so that means an average of 3–4 frames/year. The use of foundation is thought to be an additional disincentive to swarming by simulating brood nest immaturity and diverting bees to the task of wax making.

However, in order to be successful (get combs drawn quickly) and not damage existing brood by chilling, the introduction of foundation must be done at the right time and in the right position in the hive. Early in the season, in order to maintain brood nest integrity, foundation must be introduced on the edge of the brood nest – so that it becomes the next comb to be drawn if the brood nest is to expand. Later, when the colony is crammed with bees, foundation can be interleaved with frames containing brood.

Never put foundation next to the hive wall because in this position it will only be drawn as a last resort – and then usually badly.

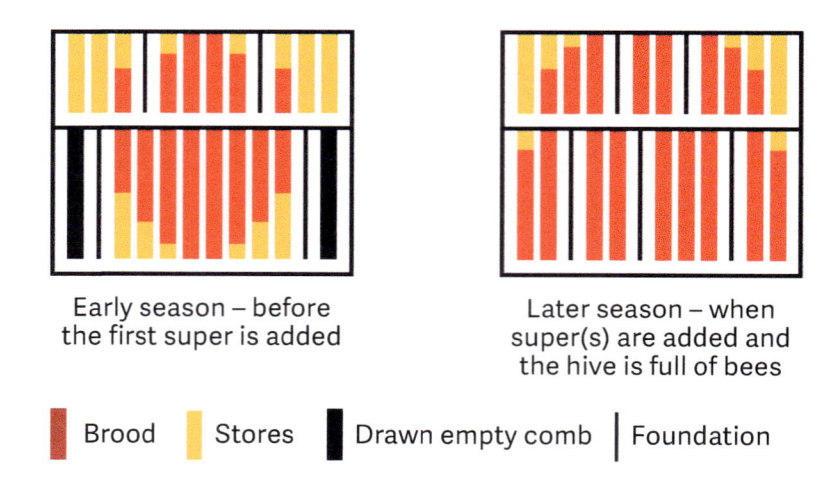

Early season – before the first super is added

Later season – when super(s) are added and the hive is full of bees

■ Brood   ■ Stores   ■ Drawn empty comb | Foundation

Figure 5   *Positioning of foundation in a hive on brood-and-a-half*

Figure 5 illustrates where foundation should be placed early in the season (left) and later in the season (right) in both deep and shallow brood boxes. Using a two-box system (ie, brood-and-a-half or double brood) the placement of foundation in the upper box is much less critical because of the warmth coming up from below. Those using a single box (standard deep or extra deep 14x12) should follow the procedure as shown for the lower box.

When introducing foundation to a hive, the beekeeper should understand that the colony MUST have an immediate need

PRE-EMPTIVE SWARM CONTROL

for more comb, either to extend the brood nest or for storage, because a colony does not engage in 'speculative' comb building. There MUST also be a nectar flow (or the beekeeper must provide one by feeding) because bees do not use stored honey for wax-making. For further information see the Welsh Beekeepers' Association booklet, *Comb Management* on the Association's website: www.wbka.com.

An alternative (less radical) approach to increasing the number of combs available for laying by the queen in the half brood is to move individual frames up into the first super.

When there is a good nectar flow early in the season, the colony will often use any available cells in the half brood to store nectar which, particularly when it has been capped, becomes a permanent fixture and will prevent their use for brood for the rest of the season. This can be rectified by moving such frames up into the first super and replacing them with frames of foundation interleaved with frames that contain brood (see Figure 6).

Frames that contain the remains of winter stores, which may be substantially sugar syrup, should be removed altogether and used to feed other colonies in need. This is to prevent contamination of the current year's honey crop with syrup.

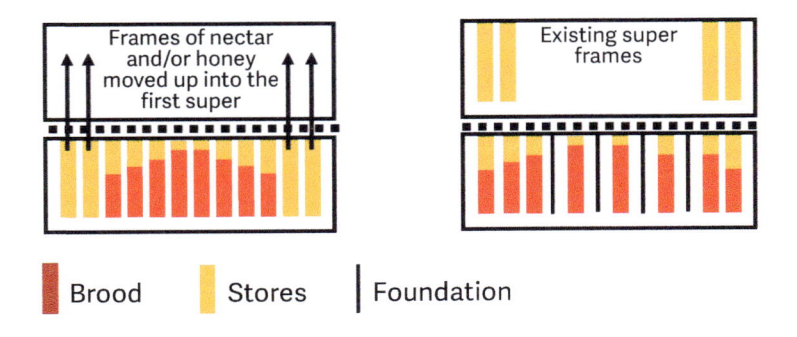

Figure 6   An alternative method of maximising the use of the shallow brood box

SWARMING BIOLOGY AND CONTROL

Having been walked on by the bees for 6–7 months, such frames can be easily identified by having dark-coloured (dirty) cappings, as opposed to fresh white or pale yellow ones.

2.2.2 Box management

This is only applicable to beekeepers who use a two-box system; either brood-and-a-half or double brood. As with comb management, the aim is to have as much of the brood nest as possible in contact with the queen excluder, thus avoiding a honey ceiling in the upper box.

A colony on brood-and-a-half

The need for box management depends crucially on where in the hive the brood nest develops at the beginning of the season. If the nest is high then nothing needs to be done because the queen is free to lay down and use as much of the available space as she needs (Figure 7a) and when a honey ceiling is created it will be in the right place – the first super.

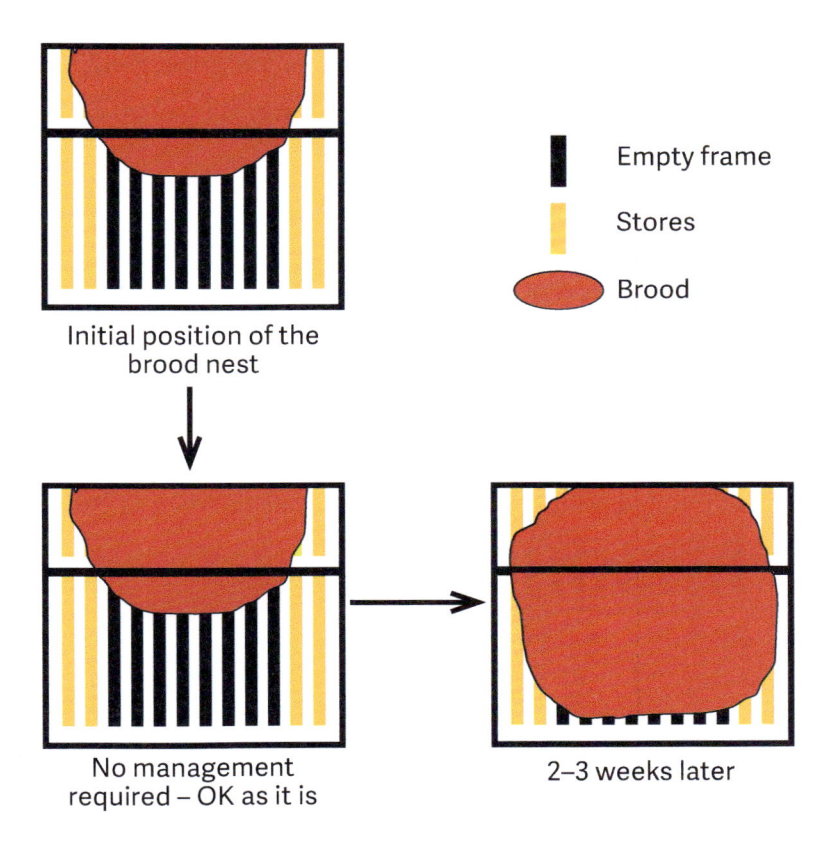

Figure 7a The initial brood nest is in a high position mostly in a shallow brood box – no management required

RESULT: *the brood nest naturally extends down to occupy as much of the deep box as required*

A nest developing in the middle of the hive (between the two boxes) will, as the season progresses, result in a honey ceiling in the upper box and this may restrict the ultimate size of the brood nest. In this situation the boxes can be swapped – placing the shallow brood box beneath the deep (Figure 7b). The only time a colony uses comb beneath the brood nest for storage is when an abundant nectar flow is in progress and then it is only temporary. Bees do not store honey in this position and as soon as they have some down time, eg, a wet

SWARMING BIOLOGY AND CONTROL

day, any honey will be removed and stored in the supers. This process can be accelerated by uncapping any sealed honey. It looks messy but it will be quickly tidied up.

When the brood nest starts low then the upper box will quickly become a honey store (a super) and the brood nest size will be restricted. In this case the best option is to place the existing shallow brood above the queen excluder (making sure the

Initial position of the brood nest

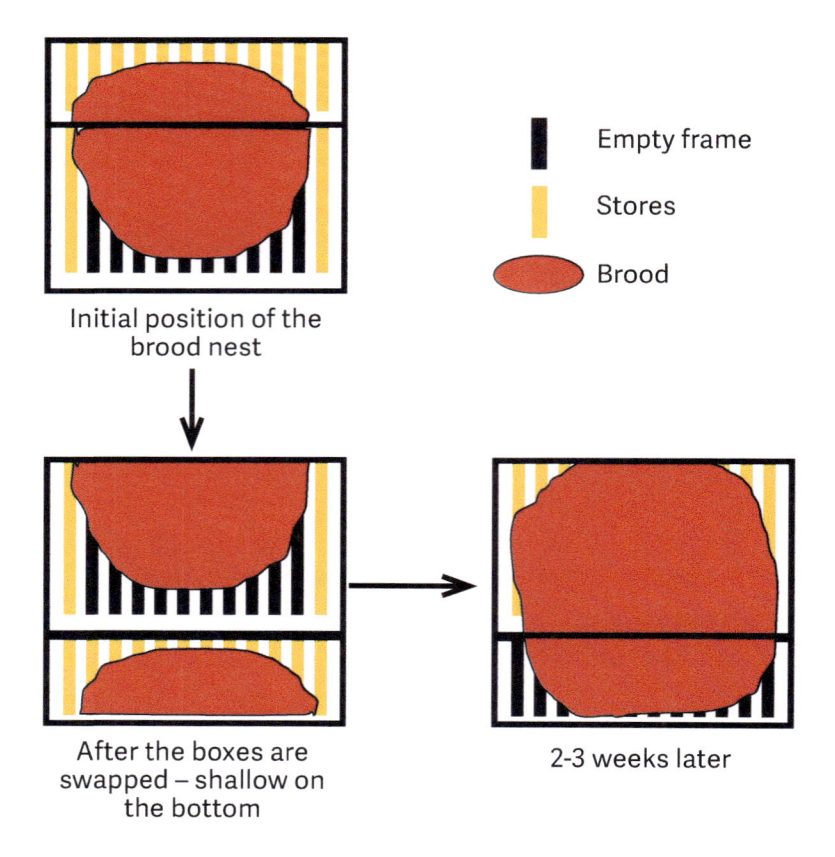

Empty frame
Stores
Brood

After the boxes are swapped – shallow on the bottom

2-3 weeks later

Figure 7b The brood nest in mid-position between deep and shallow brood boxes – reposition the shallow box under the deep box

RESULT: any honey in the shallow box is removed, comb becomes available to the queen and the integrity of the brood nest is quickly re-established

PRE-EMPTIVE SWARM CONTROL

queen is not in it) and introduce a new shallow brood box containing drawn frames beneath the deep brood (Figure 7c). If drawn comb is not available, then placing a box of foundation under the existing brood nest is the last place it will get drawn. In fact, it is unlikely to be drawn before the colony has taken the decision to start queen cells.

Initial position of the brood nest

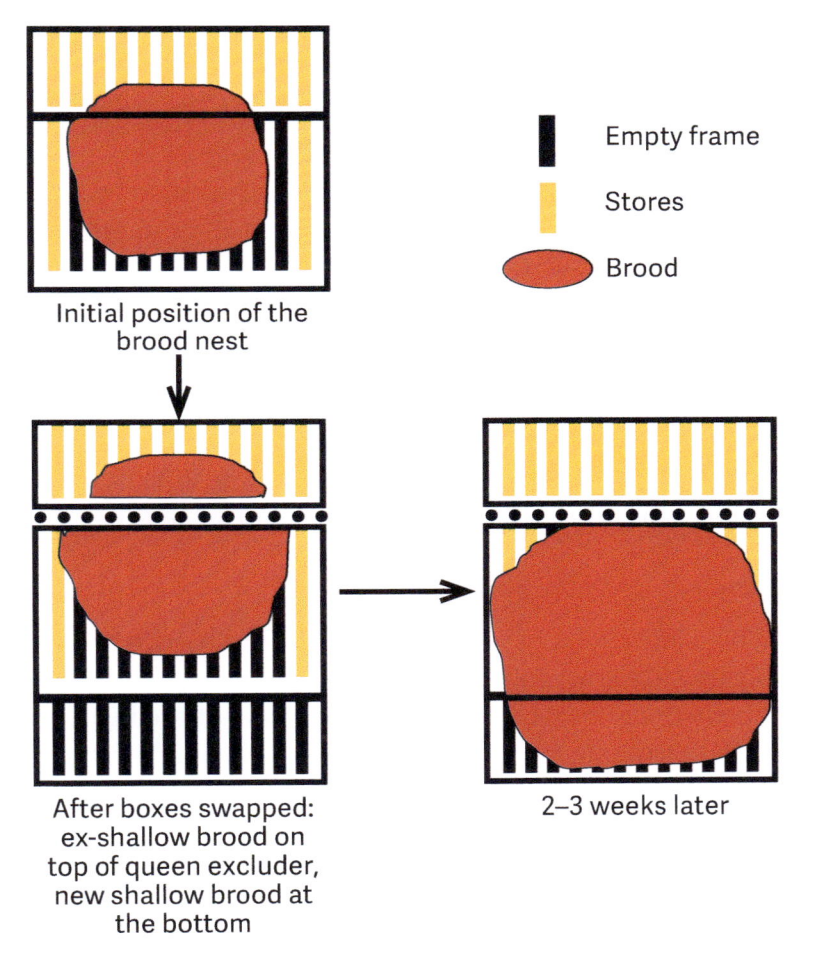

Empty frame
Stores
Brood

After boxes swapped: ex-shallow brood on top of queen excluder, new shallow brood at the bottom

2–3 weeks later

Figure 7c *The brood nest in a low position, mostly in the deep brood box*

RESULT: brood in the shallow box emerges, it becomes the first honey super and the brood nest extends down into the new shallow box below

The only option is to put the box of foundation on top of the deep brood box and below the queen excluder where it will readily be drawn. Then, before it is has accumulated a significant amount of sealed honey, move it to the bottom position. In a good nectar flow this could happen quite quickly and the beekeeper needs to be alert to catch the moment.

Creating space for the queen to lay

Contrary to what is said in most beekeeping books, the best place to create space for the queen to lay is below the existing brood nest. A box of drawn comb placed on top of the brood nest will almost inevitably be treated like a super and substantially filled with honey.

This is because bees would normally choose a smaller cavity (a volume of about 40 litres) than that provided by the beekeeper in which to build their nest. Their aim is to produce brood at the bottom and honey at the top of this space. As discussed in Part 1, when this volume has been filled it is interpreted as the signal that it is time to start queen cells and swarm.

Caution

If the hive floor you are using has a traditional depth of 21–22 mm (solid or open-mesh) it will usually be necessary to remove brace comb from the bottom bars of the box immediately above the floor, otherwise it will not fit in its new position and numerous bees will be squashed. Frames must be shaken free of bees before attempting this process.

Using floors with the correct depth for open-mesh floors (9 mm for bottom bee-space hives and 15 mm for top bee-space hives), the frames should have little or no brace comb and will not need cleaning off. This precaution only applies to bottom bee-space hives (like the National). For top bee-space hives (like the Langstroth), the traditional floor depth does not create this problem because the bottom bars of the frames are 6–7 mm closer to the floor. [Why this inconsistency in the design of beekeeping equipment has persisted for so long, I have no idea!]

Other hive configurations

In the case of a hive on a double-deep configuration, controlling the brood nest position is simply a matter of moving as many frames of brood into the top box as possible (taking care to create a sensible nest shape that the nurse bees can cover efficiently) and moving frames of stores or empty comb down. Stores beneath the brood nest will be removed quickly and space will become available for the queen to lay.

Because double brood provides more than enough space for most colonies the situation is less critical. However, the colony is likely to end the season with a substantial amount of honey stored beneath the queen excluder – which is good or bad depending on your point of view (ie, how much you want to feed).

Honey storage below the queen excluder can, however, be reduced by the judicious use of dummy boards at the sides of the boxes. The time to introduce these is when the colony has reached its peak size and it has become apparent how many combs the queen is capable of laying.

Using a single deep brood box hive configuration (which gives less space than the potential of most queens), the position of the brood nest is less likely to present problems. Here it is just a matter of ensuring that, as far as possible, all the frames pull their weight (are available for the queen to lay in). To achieve this aim, the frames should contain a minimum amount of honey and pollen stores.

Well-managed single-deep-box hives tend to have more pollen stored in the first super, simply because there is nowhere else to put it and, by default, it becomes a brood-less extension of the brood nest. Apart from difficulties during extraction

SWARMING BIOLOGY AND CONTROL

(problems balancing the frames in the extractor), pollen-clogged cells in supers tend to go hard or mouldy in storage over winter. It becomes very difficult for the bees to remove and in the process they will often create holes so that the frames have be cleaned, fitted with a sheet of foundation and re-drawn.

Extra-deep-box hives (eg, 14x12) can have a problem with brood nest position and this is not quite so easy to remedy. Figure 8 shows how this can be done by introducing frames to the middle of the box. But, because this creates a discontinuity in the brood nest, there must be a large population of bees (wall-to-wall in the box) when this is attempted.

Depending on what is available, the inserted frames can either contain foundation or empty drawn combs. Alternatively, existing lateral frames (from the outside of the box), with the cappings of any honey stores scored using an uncapping fork to encourage removal of the contents, can be used to good

■ Brood

■ Stores

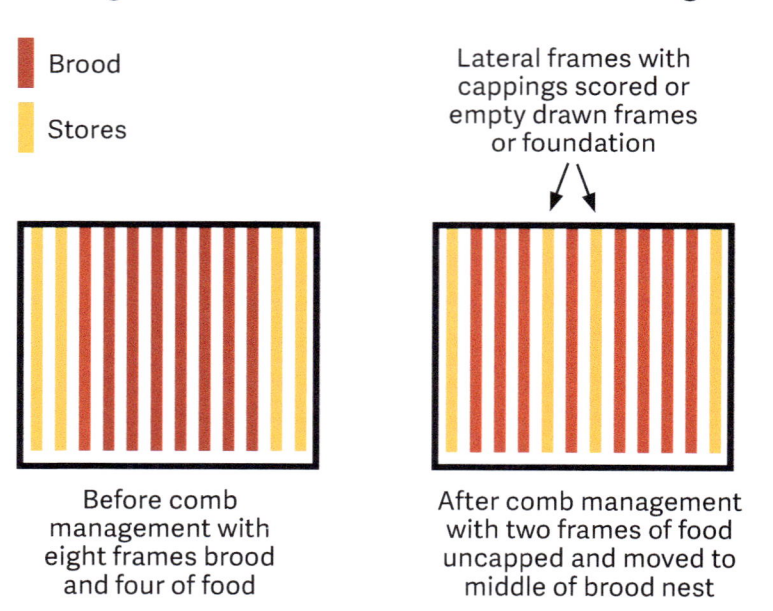

Lateral frames with cappings scored or empty drawn frames or foundation

Before comb management with eight frames brood and four of food

After comb management with two frames of food uncapped and moved to middle of brood nest

Figure 8   Expanding the brood nest in single box hives using a deep or extra deep box

PRE-EMPTIVE SWARM CONTROL

effect. The colony will immediately seek to restore brood nest integrity and these frames will quickly become part of a larger, spherical brood nest.

Supering

Also part of box management is adequate supering of a colony. It is essential to provide plenty of space for the processing of nectar and the storage of honey. It should be remembered that fresh nectar has two to four times the volume of the honey that it will become and extra space should be provided for the bees to conduct the drying process.

Supering needs to be kept one jump ahead of the need for storage but without creating so much volume that heat loss becomes a problem. It is the time to add the next super when the top super is full of **bees** (wall-to-wall), regardless of the fact that it may be only partly full of honey. However, no amount of supering will act as a substitute for poor brood nest management.

*Add the next super when the top super is full of **bees**, regardless of how full it is of honey*

Weather is the problem that the beekeeper can do nothing about and during prolonged adverse conditions the bees will inevitably move down from the supers, crowding the brood area, and this will often trigger swarming.

2.2.3 Brood relocation

This is one of the oldest tricks in the book and usually goes under the name of the Demaree method, dating back to 1892 – a method that is probably under-used in modern beekeeping. Unfortunately, it is also widely misunderstood and I have read and heard Demaree sold as a method of artificial swarming (re-active swarm control), which it definitely isn't. I have also seen it suggested as a method of making increase or raising queens, for which purpose it is likely to be very hit or miss (with the emphasis on miss).

There are many minor variations of the method but the basic principle is the removing of frames of brood from the bottom of the hive and relocating them in a new box at the top of the hive – above the supers. This box is fully connected to the main hive and should have no separate entrance which means that the whole colony is kept together and not split.

If the box of frames containing brood is placed on a split board and separated from the colony below, it is something else entirely and not a Demaree. Using the Demaree method, bees wishing to exit the hive must pass down through the supers and through the brood area at the bottom of the hive. This is where they will usually have enough contact with queen substance to regard themselves as being in a queen-right colony.

However, if there are a lot of supers on the hive and vertical separation from the brood area is considerable (3+ supers), there may not be enough queen substance at the top of the hive (which can only be transferred through vertical traffic of bees). If this happens, the bees up there may regard themselves as being queen-less and start emergency queen cells.

At the bottom of the hive, the removed frames are replaced by empty drawn comb (if not available, foundation can be used) thus giving extra space for the queen to lay. Brood at the top of the hive attracts nurse bees to move up to cover it and this serves to reduce congestion at the bottom of the hive. It may also slightly reduce the queen's laying rate if it creates a temporary shortage of nurse bees.

The combination of new laying space for the queen and a reduction in congestion in the brood area is what inhibits the impulse to swarm (start queen cells).

The method was originally designed for very prolific bees which required double (or triple) deep brood boxes. When brood has emerged from the frames moved to the top of the hive, the plan is to return them to the bottom of the hive in exchange for ones containing more recent brood (a frame circulation system). However, this is not usually possible (see the discussion of downsides below).

Figure 9a shows a classical Demaree with two deep brood boxes. Figure 9b shows how the method can be applied to a hive on a single deep brood box. The manipulations shown in Figure 9b can easily be adapted to a brood-and-a-half configuration – the shallow brood stays at the bottom of the hive, either above or below the deep brood box depending on the position of the brood nest (see Section 2.2.2 above).

When managing a hive on a single deep brood box, it is doubtful if the beekeeper would ever want to relocate the entire brood nest to the top of the hive as this would seriously unbalance the colony. Typically only 4–8 frames are moved at any one time, resulting in an incomplete Demaree box, as shown in Figure 9b. An incomplete set of frames should be flanked by a dummy board on either side (colour-coded blue).

Further frames of brood can be moved up at a later date if required so it may eventually become a complete box.

SWARMING BIOLOGY AND CONTROL

A part-filled Demaree box is no problem but the beekeeper needs to be aware that in an abundant nectar flow wild comb may be built in any unoccupied space at the sides of the box.

Downsides

The Demaree method is quite an effective method of pre-emptive swarm control but it does have some downsides.

The first is that the bees covering the brood at the top of the hive may be far enough removed from the queen that they regard themselves as queen-less and start emergency queen cells. The greater the spatial separation (the number of supers) the more likely this is to happen.

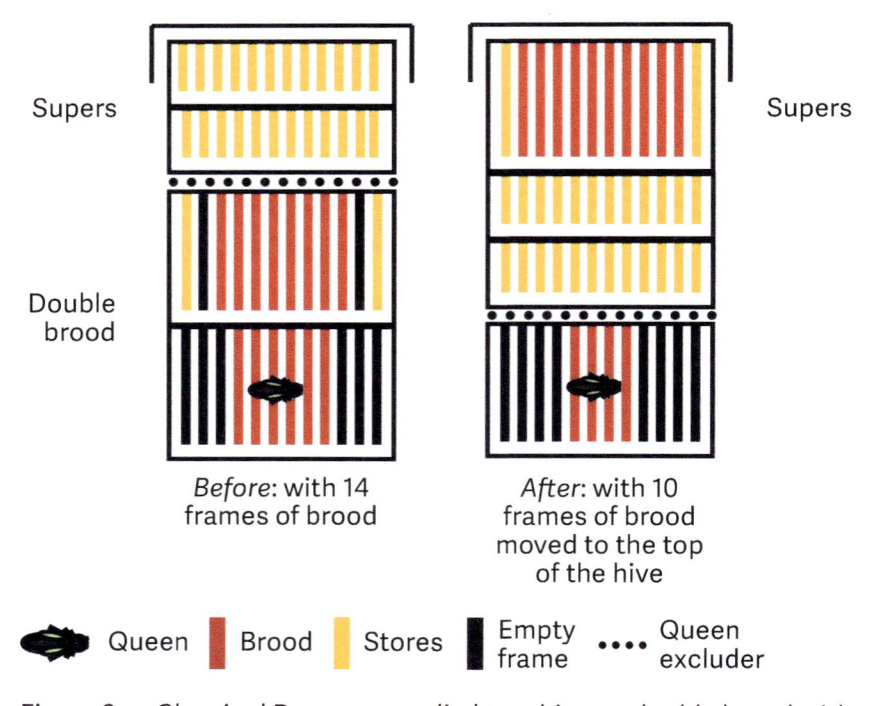

Before: with 14 frames of brood

After: with 10 frames of brood moved to the top of the hive

Queen | Brood | Stores | Empty frame | •••• Queen excluder

Figure 9a   Classical Demaree applied to a hive on double brood with 14 deep frames of brood
NB. For really big colonies a double brood can be retained at the bottom

PRE-EMPTIVE SWARM CONTROL

After five to seven days it is necessary to carefully examine the top box for queen cells and destroy them. Without a separate entrance, a virgin queen emerging in the Demaree box will not be able to mate unless she moves down through the supers and manages to squeeze through the queen excluder. If this happens it is possible she will kill the existing queen (it's either kill or be killed!) so it is better to avoid this problem.

The second is that frame recycling is more difficult than it appears at first sight because as soon as the brood has

Before: with a single deep brood and 12 frames of brood

After: with eight frames of brood moved to top and four remaining at the bottom

Supers

Box with an incomplete set of frames flanked by dummy boards

Queen | Brood | Stores | Empty frame | Dummy board | Queen excluder

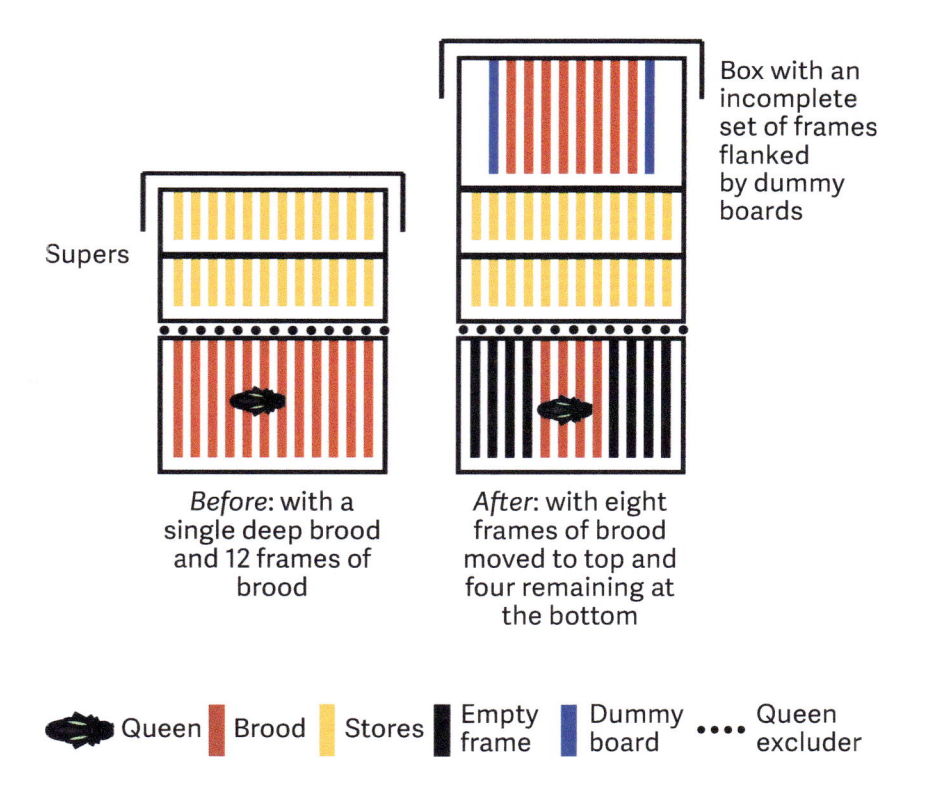

Figure 9b   The Demaree method applied to a colony on a single deep brood box with 12 frames of brood

NB. The colony could have been on brood-and-a-half. The shallow brood would have remained at the bottom, either over or under the deep brood box emerged in the Demaree box, the vacated cells are quickly filled with nectar and later with capped honey. This means that the Demaree box normally has to be left in place for the rest of the season and removed as part of the honey harvest. However, providing the combs are fairly new, this should not adversely affect honey quality.

2.2.4 Splitting colonies

The previous methods of pre-emptive swarm control have kept the colony in one piece. Splitting is different and (potentially) creates a second colony. However, splitting is the most powerful and reliable method of pre-emptive control and has a long history of use in beekeeping.

Split boards came into use in the latter part of the nineteenth century and there have been many different designs over the years, culminating in the 'all singing, all dancing' Snelgrove board. Splitting colonies is also a method of making increase, which is covered in more detail in the WBKA booklet, *Simple Methods of Making Increase* on the Association's website, www.wbka.com.

It is often said that splitting a colony is the enemy of a good honey crop. However, splitting is always better than losing a swarm – unless you can guarantee catching it!

The effect on the honey crop depends crucially on the timing of the split. If it is done at the right time, eg, directly after the spring flow, it can result in an enhanced yield. Under the right circumstances, the two resultant colonies can produce more than the original (one) colony – even if it did not eventually swarm. A controlled split is always better than an artificial swarm because it enables the beekeeper to create a better (more functional) age-class distribution of worker bees in both parts of the split.

When splitting a colony to provide pre-emptive swarm control the following considerations should be taken into account:

PRE-EMPTIVE SWARM CONTROL

- the split should be sufficiently radical to provide swarm control for the rest of the season
- both sides of the split should be viable, ie, have adequate bees, brood and stores
- the timing should be right for the colony, ie, with regard to its state of development
- the timing should be right for potential nectar flows, ie, allowing time for the colonies to rebuild.

Timing

The most common question asked about splitting hives is, 'When should I do it?' The simple (and not very helpful) answer is, 'When it's ready'. No firm guidance (prescription) is possible and the details of splitting (both timing and the allocation of resources) depend on the judgement of the beekeeper. Basically, the colony has to be large enough to withstand splitting and for both halves to have sufficient assets in terms of bees and brood (and stores) to enable them to rebuild without any undue check.

A colony on a brood-and-a-half needs to have a minimum of eight good frames of brood in the deep box and six in the shallow. More is better, of course, but what you must remember is that you are walking a tightrope between a well-developed colony and one that will set up to swarm in the near future. Waiting that extra week for the colony to develop further can result in the colony making its mind up for itself.

Again this is a matter of judgement and that can only be gained by experience (some of it hard!).

Figure 10a (using a new hive stand) and Figure 10b (using a split board) are examples which illustrate the principle. In both figures, the blue box and the frames within are new.

Recombining splits

If a split has been made properly (see guidelines above) there

SWARMING BIOLOGY AND CONTROL

Before: colony on
brood-and-a-half +
two supers

After: Seven deep
frames of brood
removed, eight
shallow frames of
brood remain in place

Queen cells produced

New box with seven
deep frames of brood

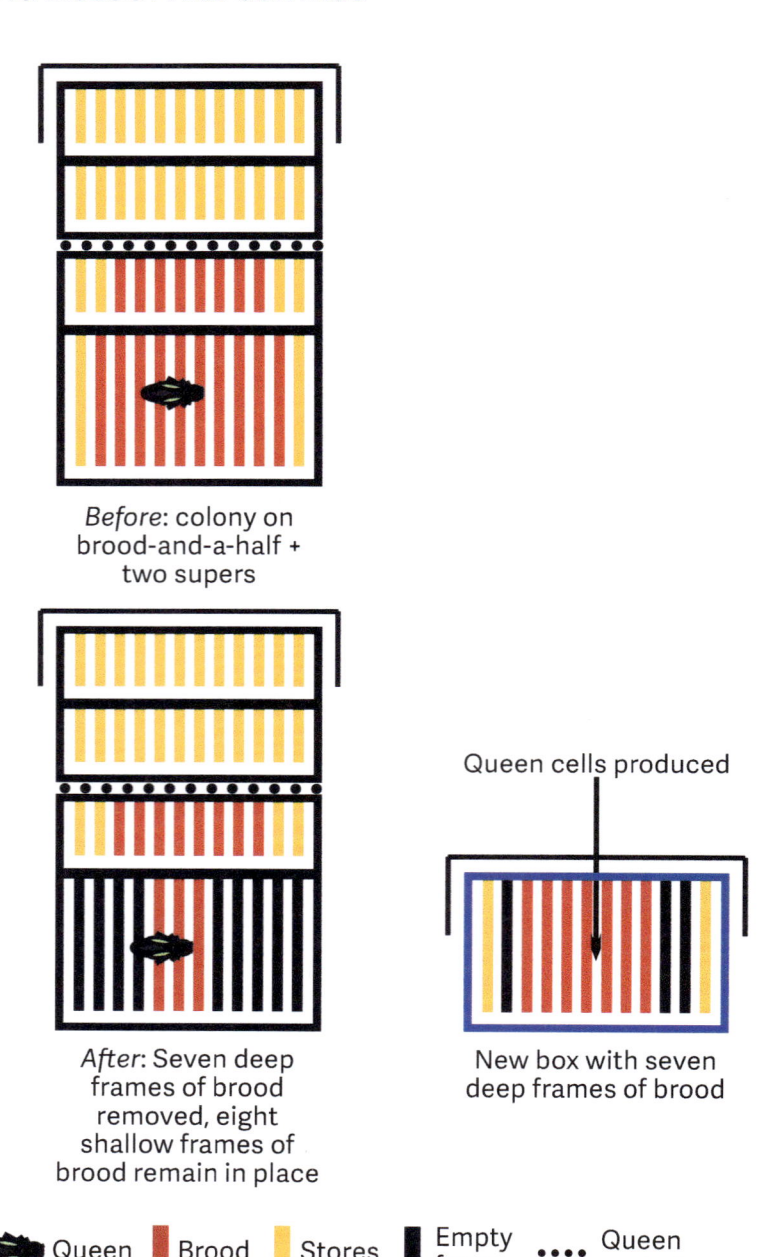

■ Queen ▌Brood ▌Stores ▌Empty frame •••• Queen excluder

Figure 10a Splitting a colony on brood-and-a-half with 10 deep frames of brood, seven of which were placed in a new box on a new hive stand

PRE-EMPTIVE SWARM CONTROL

Shallow +10 frames of brood →

Deep +10 frames of brood →

Queen cells made

Before: colony on brood-and-a-half with two supers

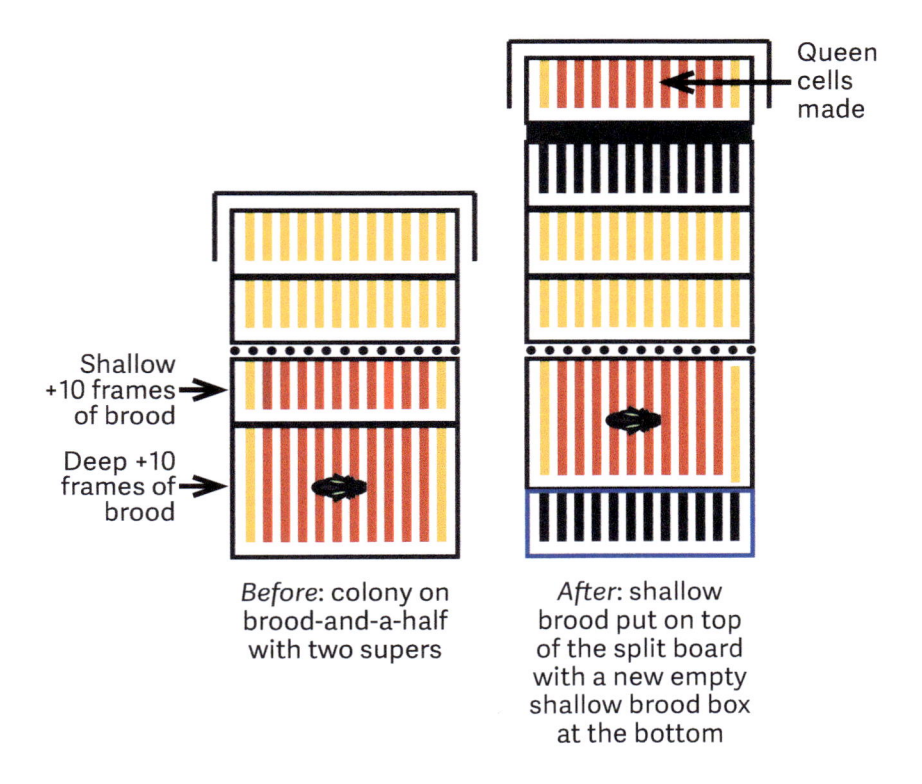

After: shallow brood put on top of the split board with a new empty shallow brood box at the bottom

Figure 10b Splitting a colony on a brood-and-a-half by placing the entire shallow box over a split board

is about a 90+% probability that the queen-less part of the split will raise a new laying queen. The beekeeper now has two colonies instead of one and this may not be the desired outcome. Simple logic dictates that you cannot go on doubling the number of your colonies indefinitely so the solution is recombination (uniting colonies), but how and when?

Recombination early in the season is not usually a sensible option as it will only create a large colony that is likely to undo all your good work and try to swarm. Later in the season, when the swarming impulse is on the decline, it is possible to recombine to produce a 'super-colony' for the main honey flow or for taking to the heather.

SWARMING BIOLOGY AND CONTROL

There are three main options for the new colony (with a new queen) which will currently be living either on a new stand or a split board.

- Giving it permanent independence, ie, creating a new long-term colony.
- Recombining later in the season to produce a 'super-colony' (with the choice of queen to be made by the beekeeper).
- Giving it temporary independence, ie, supering it, keeping it until the end of the season and then recombining (again with a choice of queen).

The word 'recombination' implies that the new colony has to be united with the original colony. However, much greater flexibility is possible and it is often the preferred option to paper-unite it with a colony that has an old or unsatisfactory queen as a simple method of re-queening.

Artificial swarming also creates two colonies from one and there is further discussion about how to deal with this situation in Chapter 2.4 (Aftermath of Artificial Swarming).

Uniting over paper is often the preferred option

2.3 RE-ACTIVE SWARM CONTROL (PRELIMINARY INVESTIGATION AND DIAGNOSTIC TREE)

You have opened a hive and found queen cells!

The knee-jerk reaction (more of a hand-jerk really) of many beekeepers is to destroy them under the misguided notion that this will prevent swarming. So the first rule is 'get a grip' and DON'T PANIC! Destroying queen cells never has been and never will be a successful method of swarm control.

If you destroy one lot of queen cells, the colony will immediately make some more and will probably swarm earlier than normal in their development – often before the first queen cells are sealed. If you destroy queen cells twice you run the risk of the colony swarming and leaving behind no provision for a new queen. Any delay in swarming that you induce by destroying cells will often result in the prime swarm being larger than it would have been if you had not interfered.

Once a colony of bees is triggered to swarm it is very rare for them to go off the idea.

Taking control

At this point the beekeeper must take control of the situation by using an appropriate form of management.

For an attentive beekeeper (who does regular inspections to check for signs of swarming), the queen cells will usually be unsealed (or possibly some newly sealed) so this will almost certainly be before the colony has issued a prime swarm. In these circumstances making an artificial swarm is the solution.

If the colony has already swarmed then other management techniques can be used to prevent a cast swarm and further loss of bees, but the details depend on the current status of the colony. Approaching the problem logically and finding out exactly what stage of the swarming process has been reached will give the beekeeper the best chance of successfully intervening, thereby not losing bees, saving as much of the potential honey crop as possible and not ending up with a queen-less colony.

Just occasionally, queen cells are torn down and the colony seems to go off the idea of swarming. It is said that this can happen naturally if an abundant flow of nectar occurs or that it can be artificially induced by removing several frames of brood and replacing them with foundation. The intervention of a nectar flow is outside the beekeeper's control and the 'shock' removal of brood and the introduction of foundation may occasionally work but can certainly not be relied upon.

Colony diagnosis

The latter part of this chapter consists of a chronological diagnostic tree consisting of 12 identifiable steps covering the process of swarming from the beginning to the end – in this case the 'bitter end' of finding the colony queen-less or having a drone-laying queen.

RE-ACTIVE SWARM CONTROL
(PRELIMINARY INVESTIGATION AND DIAGNOSTIC TREE)

Before you contemplate any management of a colony that has developed queen cells, you first need to understand what is going on. For example, are the queen cells really there for swarming or is there some other reason? If it is swarming, what stage in the process has the colony reached?

All the information you need to make this diagnosis is 'written' on the brood combs and, to a lesser extent, the bees. But before you make your diagnosis you have to know how to 'read' the combs – understanding what you can see and what else you need to look for in order to reach an accurate diagnosis. In many ways this is like a forensic investigation. To be able to do this effectively you must have a basic knowledge of brood development, honey bee biology and behaviour. The diagnostic tree is designed to help the beekeeper through this process and arrive at a correct decision concerning management.

Reasons for queen cells being present in a hive

The different types of queen cell, why they are there and the way the colony behaves in their presence has already been covered in detail in Chapter 1.4 but just to summarise these are:

- reproduction – swarming
- replacement of the existing queen – supersedure
- the colony is queen-less – emergency re-queening.

These are three separate and distinct behavioural programmes but nine times out of ten (or probably nearer 19 times out of 20), when a beekeeper opens a hive and sees queen cells it is the swarming programme that the colony is following.

We often refer to the 'types' of queen cell because the origin of the cell and the larva that is receiving a special diet to transform it into a queen, its position in the hive and the number of cells that have been produced are characteristic of the programme that has been activated. These are important clues as to what is going on but not completely definitive (see 'Ambiguous situations' below).

SWARMING BIOLOGY AND CONTROL

Before deciding on any management in response to finding queen cells it is necessary to correctly identify which programme the colony is on. The photos below show typical queen cells (external appearance and position) for the three programmes. Only the presence of swarm cells means that the colony is intent on swarming and those produced for other reasons (supersedure and emergency re-queening) will on NO account result in the issue of a swarm. Supersedure and emergency queen cells do not usually require any intervention from the beekeeper – except to leave the bees well alone and let them get on with it.

So, how do you know which programme the colony is following?

Swarm cells  *Supersedure cell*  *Emergency cells*

Ambiguous situations

In the vast majority of cases when a beekeeper opens a hive and finds a significant number of queen cells the intentions of the colony are completely obvious – it is SWARMING and there is no room for confusion. However, it should be emphasised that it is not the appearance of queen cells or their position on the frame that is most important, it is why they are there in the first place; in other words, the behavioural programme that the colony is following.

RE-ACTIVE SWARM CONTROL
(PRELIMINARY INVESTIGATION AND DIAGNOSTIC TREE)

Swarming or supersedure?

If there is a limited number of queen cells present the question arises, is it swarming or supersedure? It is not always obvious. For example, swarm cells are not always at the edges of frames and supersedure cells are not always on the face of the combs and, just to confuse matters, the number of cells may be atypical – too few to be sure about swarming or too many to be sure about supersedure.

Fortunately, emergency re-queening is always obvious because there will be no eggs in the colony and the youngest brood will tell you exactly when the queen was lost and the queen cells were started.

So the most common problem is distinguishing between swarming and supersedure. In swarming there are usually multiple queen cells present but what is happening if you find just three or four queen cells by the bottom bar of a frame – which programme is this?

If the cells are on more than one frame it is more likely to be swarming but if they are on the same frame it is probably supersedure – but you can't always be sure. Similarly, more than three queen cells on the face of a comb is usually regarded as too many cells to be a supersedure – but still it may be!

If the queen cells are newly started and will not be sealed for several days (and there is no immediate risk of a swarm) the decision as to what is going on can be deferred. The beekeeper can come back in two to three days' time and if more queen cells have been started it is almost certain that swarming is the intention. If there are no additional queen cells then it is probably a supersedure. If the cells are sealed (or close to sealing) and swarming is probably imminent and you still can't be sure, it is best to play safe. A belt and braces method of dealing with this situation is described in Appendix 1.

SWARMING BIOLOGY AND CONTROL

Swarm and emergency cells together

There are some circumstances where swarm and emergency (type) queen cells can coexist in a colony. For example, if a colony swarms early (before the swarm cells are sealed), the workers will usually respond to what they perceive as loss of the queen by making some additional emergency cells. This is a direct response to the sudden loss of queen pheromones and the bees that are responsible seem to be unaware that they already have numerous swarm cells present. The same thing can happen if the beekeeper removes the queen when making an artificial swarm.

In neither of these cases has the colony changed to an emergency re-queening programme and the emergency cells may or may not be of practical importance later, depending on subsequent management. However, there is one situation where emergency queen cells can matter and that is when a colony has already swarmed and the beekeeper has been forced into culling all but one of the remaining queen cells to prevent the issue of a cast swarm.

If the culling is done immediately after the prime swarm has departed there will still be eggs and young larvae present from which the bees can make emergency queen cells. These cells will be started after the beekeeper has closed up the hive with a glow of self satisfaction that everything is under control. Being in swarming mode, the colony may proceed to cast swarm with the queen that emerges from the queen cell the beekeeper left intact, treating the emergency cells as backup to provide their new queen.

Again it is not the type of cell that matters but the behavioural programme the colony is on that determines the outcome.

Some other basic facts you need to know

The developmental stages of the three types of brood (queen, worker and drone) and timing (in days) are shown

RE-ACTIVE SWARM CONTROL
(PRELIMINARY INVESTIGATION AND DIAGNOSTIC TREE)

diagrammatically overleaf. In order to understand what you are looking at in the hive you need to familiarise yourself with the key developmental stages, their appearance and their timings.

Queen cell development

The earliest you can identify a viable queen cell is when it is already three days old – an egg in a queen cup does not necessarily mean it will become a queen cell (see Chapter 1.2). The critical decision for the colony is made when the egg hatches out (Day 3) and the nurse bees start to feed the larva with royal jelly. A queen cup with a pool of royal jelly and a tiny larva (which is almost impossible to see) in it will almost inevitably be taken to full term and become a sealed queen cell. Sealing takes place on Day 8, ie, the larval feeding period is just five days.

Once queen cells are sealed it is difficult to know how old they are without breaking a few open to inspect their contents. There are often cells present covering a range of ages, so if you want to know the exact timing you need to look at cells in several different parts of the hive to ensure you have covered all eventualities. In some instances the age difference between the oldest and youngest queen cells can be in excess of seven days.

Emergence of the queen occurs on Day 16, ie, eight days after sealing. A queen cell from which the queen has recently emerged usually has a hinged lid attached (sometimes it may fall off) but it is also quite common for the bees to close the lid and reseal it – look for a line round the tip of the cell (a sort of 'tear here' line). You may be surprised to find an occupant in such cells but usually it is a worker bee that has gone in to do a bit of cleaning work and has been sealed in by some tidy minded sister. If the bee is head-down in the cell it will be a worker but, if it is head-up, it will be a queen and she will just be waiting for you to open the cell so she can walk out and greet the world. Again, do not panic and kill her! Simply let

SWARMING BIOLOGY AND CONTROL

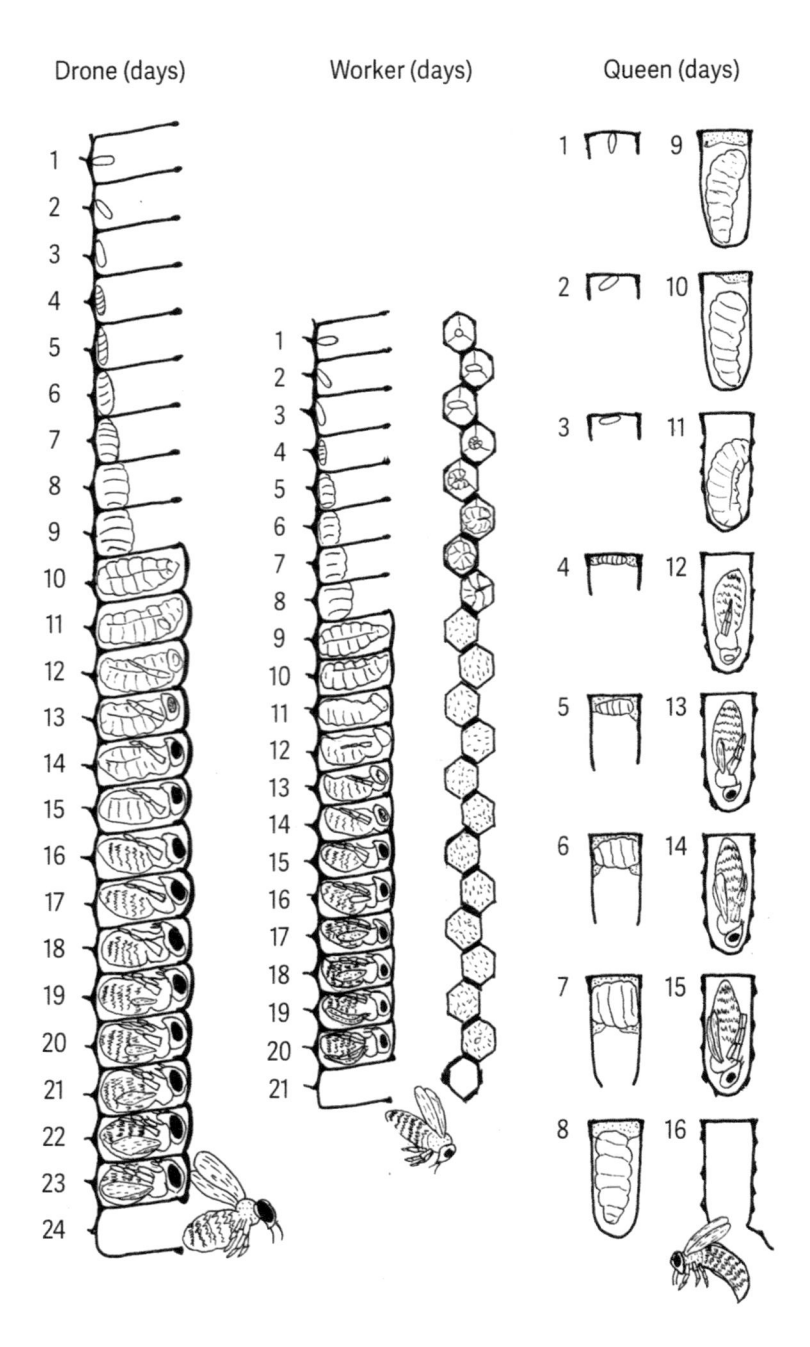

The developmental cycle for drones, workers and queens

RE-ACTIVE SWARM CONTROL
(PRELIMINARY INVESTIGATION AND DIAGNOSTIC TREE)

Letting a virgin queen out into the colony

her walk off into the colony because this is an extremely easy (even advantageous) situation to resolve (see Step 7 below).

Worker brood development

If when you open the hive you find sealed queen cells (or those on the point of being sealed), the colony may already have swarmed. You may already have some clue that this has happened from the number of bees in the hive being fewer than you expected. Confirmation that the colony has not yet swarmed is finding newly laid eggs (standing on end in the bottom of the cells) or, better still, seeing the queen herself. If there are only eggs lying down in the cells then this is inconclusive and you really need to see her (the queen) to be absolutely sure. If there are no eggs then the colony has almost certainly swarmed and the age of the youngest larvae will tell you when this happy event (happy for the bees but less so for you) occurred.

SWARMING BIOLOGY AND CONTROL

Recently laid eggs standing upright indicate the colony has not yet swarmed but it is better if you can see the queen

Eggs hatch on Day 3. Worker brood is sealed on Day 9 and emerges on Day 21, so there are six days of feeding and the bigger the larva, the older it is. The reason you need to know when the colony swarmed is so that you can assess how imminent is a cast swarm. It is also useful to know when the swarming actually took place so that you understand how you managed to miss seeing the warning signs – this is an important learning opportunity!

If there are no unsealed larvae in the colony – only sealed brood – then it is at least nine days since the colony swarmed and you are in serious trouble because a cast swarm is imminent or has already departed. For how best to recover from this situation, see Steps 4 and 5 below.

Drone brood development

Drone brood is sealed on Day 10 and emerges on Day 24 or even as late as Day 28. There is not much to be said about drone brood except that you should not rely too much on the information that can be gleaned from its stage of development. In a swarming situation, drone brood is the first to be neglected by the worker bees; it may be poorly fed, remain unsealed for long periods and even die. Even unsealed worker brood is subject to a higher mortality rate in a queen-less, post-swarming colony.

Chronological diagnostic tree and remedial management

What follows is a diagnostic tree designed to assist the beekeeper. The main period for swarming is May to July, peaking in late May and throughout most of June. Outside this period swarming is less likely but can still happen. It is possible for a colony to go from no obvious signs to actually issuing a swarm in five days or less, ie, before any queen cells are sealed. In order to catch swarming in the early stages, regular hive inspections are required but how often should these be done? It is suggested that a seven-day inspection frequency will probably suffice but it is not completely foolproof.

A well-known method of extending the period needed between checks for queen cells to 14 days is to clip the queen's wings, thus preventing her from being able to fly far. When a colony with a clipped queen tries to swarm, the queen is not airworthy and falls to the ground. The bees cluster around her for a while but eventually return to their hive in disgust and await the emergence of a virgin queen with which they will swarm.

Knowing that the queen has clipped wings should enable the beekeeper immediately to identify what has happened and apply the appropriate management to prevent cast swarming (see Step 5 or Step 7 below – your choice as to method). Occasionally, failure of the prime swarm can also occur naturally when, for some reason, an intact queen is unable (or unwilling) to fly. To identify this situation and prevent the colony re-swarming with a virgin queen requires remedial management (see Appendix 4).

SWARMING BIOLOGY AND CONTROL

Quick inspection for queen cells

One of the advantages of keeping bees on a two-box system (brood-and-a-half or double brood) is that (swarm) queen cells will usually be started along the bottom of the combs of the upper box. If no other management is required, a swarm check can be accomplished very quickly by simply lifting one side of the upper box and looking for signs of queen cells by the bottom bars (using smoke to move the bees and get a proper look).

This type of inspection is not 100% reliable but good enough for most purposes. It saves time, reduces disturbance for the colony and is certainly a lot better than not bothering to look at all! Single brood box systems will always require the removal of at least some frames to check for queen cells.

A colony intent on swarming – and how!

Regular inspections (swarm checks) will ensure that you never need to use the later steps in the diagnostic tree that follows. Steps 1–3 cover pre-swarming development, with Step 3 putting the beekeeper on amber alert. Steps 4–9 deal with increasingly more advanced stages in the swarming process

RE-ACTIVE SWARM CONTROL
(PRELIMINARY INVESTIGATION AND DIAGNOSTIC TREE)

and Steps 10–12 deal with problems that may arise after the swarming process is over and the colony has not returned to normal with a laying queen – in other words, a rescue programme.

Each step consists of two parts. *Investigation* – instructions to help you to identify what stage the colony is at – and *Remedial Action* – what to do about it (management required).

STEP 1 – There is drone brood in my colony

Investigation

None necessary. This step is included because it is a widely held myth that the presence of drone brood means that the colony is preparing to swarm.

The presence of drone brood is merely an indication that the colony has reached a certain stage in its spring build-up when it can 'afford' to produce and support drones. This is

Drone brood

also another way in which the colony is able to pass on its genes (by a drone getting lucky and mating with a virgin queen from another colony), so it may as well avail itself of this (rather low probability) route as early in the season as possible.

All healthy, successful colonies produce drones as part of their normal development, usually starting in mid-March and continuing until sometime in August. Many of these colonies with early drone brood will make no attempt to swarm during the season.

Remedial Action

No action is required. Just rejoice that the colony is developing normally but be aware that the presence of drone brood means

SWARMING BIOLOGY AND CONTROL

that the varroa mite population will start to grow more quickly and now is the time to check that the colony does not have a mite problem that will only escalate as the season progresses.

STEP 2 – There are queen cups in my colony

Investigation

Check to see there are no queen cup contents; no eggs and particularly no young larvae in pools of royal jelly.

Remedial Action

If there are no cup contents, no action is required. Like drones, the building of queen cups (practice cups or fun cups), mostly near the bottom bars of frames, is a natural stage in the build-up of the colony and does not mean that swarming is imminent. Cup building is thought to commence when the queen is no longer regularly walking on the edge of the frames and leaving her footprint pheromone there – presumably because she is too busy with other matters and the hive is also becoming more congested.

A queen cup

STEP 3 – There are queen cups with standing-up eggs in them (eggs standing vertically at the bottom of the cell)

Investigation

Check that no cells have gone a stage further and contain a larva in a pool of royal jelly.

Remedial Action

If there are only standing-up eggs no action is required except to go onto amber alert – this may be the start of something

RE-ACTIVE SWARM CONTROL
(PRELIMINARY INVESTIGATION AND DIAGNOSTIC TREE)

more serious. However, many colonies will have eggs in queen cups several times during the season and still make no attempt to proceed any further.

A queen cup with a standing-up egg

STEP 4 – There are queen cups with contents (larvae and royal jelly) in my colony and some of the cells are starting to be extended

Investigation

This is a sure sign that the colony is almost certainly going to swarm – so it is red alert time.

Now you need to determine what stage of development the swarm cells are at so that you can estimate how many days it will be until they are sealed and therefore the likelihood of swarming in the near future.

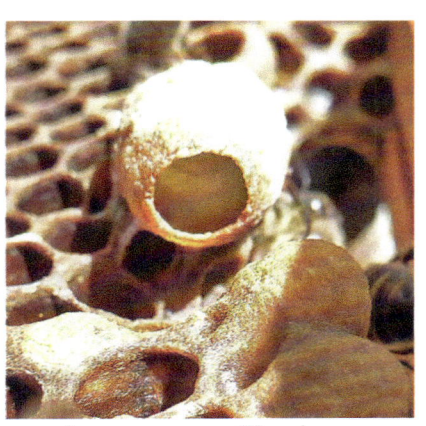

A queen cup with a larva

Just occasionally a colony will swarm prematurely – before there are any sealed swarm cells – so it is as well to check this has not already happened. Is the colony smaller than you expected? Are there newly laid eggs or, better still, can you see the queen? If you decide the colony has already swarmed you need to go to Step 5.

SWARMING BIOLOGY AND CONTROL

Remedial Action

If all the queen cells are at an early stage of development then you probably have time on your side (1–3 days?). But do not procrastinate; remember that some colonies swarm prematurely. So, in reality, you have to prepare to carry out an artificial swarm on the colony as soon as possible. There are many methods of artificial swarming to be found in beekeeping books. However, it is recommended that you use the method described in Chapter 2.4 and accompanied by colour-coded diagrams (Snelgrove II modified). Apart from its reliability, this method has the advantage that you do not have to find the queen to carry out the initial manipulation.

STEP 5 – There are sealed queen cells in my colony

Investigation

Now you must seriously consider whether or not the colony has already swarmed.

The first clue is the number of bees in the hive. Are there fewer bees than when you last looked at them? Are the supers full of bees?

Next look at the brood. Are there newly laid eggs? Can you find the queen?

Sealed queen cells

If there are no eggs, what is the youngest brood you can see? If you can find standing-up eggs or have seen the queen you are in luck and the colony has not yet swarmed – but it may do so at any moment if the weather is favourable. If it is now late afternoon (say after 16.00), seeing standing-up eggs is not completely reliable. This is because the colony could have swarmed earlier in the day so, to be absolutely sure, you will need to see the queen.

RE-ACTIVE SWARM CONTROL
(PRELIMINARY INVESTIGATION AND DIAGNOSTIC TREE)

If, after careful investigation, you decide the colony has not swarmed, go back to Step 4 and make an artificial swarm. If you decide it has swarmed then there is only one option open to you.

Remedial Action

If you conclude that the prime swarm has issued from the colony, all you can do now is prevent a cast swarm and the further loss of bees. The accepted method of doing this is to reduce the queen cells down to one so the colony has no option but to settle down with the only remaining queen that can emerge. But what rule do you follow?

There is a range of advice given on this matter. The most common advice is that you select an unsealed queen cell in which you can see a healthy larva and destroy ALL the rest (sealed and unsealed). If there are no unsealed cells you will have to settle for a sealed one – the best you can find and preferably one in a well-protected position so there is no chance of accidental damage during manipulation.

Personally, I do not see the point in the unsealed cell option. Bees will not seal queen cells when the occupant is already dead and all you are doing is foregoing the advantage of a sealed cell from which the queen will emerge sooner. Of course, some queens die in the cell after it has been sealed and there is no way of avoiding this possibility – but you must be aware it can happen and be prepared to take remedial action.

Some writers recommend taking out insurance and leaving two cells of the same age – but how do you know cells are of the same age? Using this method there is still the potential for a cast swarm. The golden rule of this recovery operation is that you MUST destroy ALL queen cells apart from the one (or two) that you select. To do this properly, it is usually necessary to gently shake or brush bees from the combs so that you can see what is there. Be particularly careful not to miss any queen

SWARMING BIOLOGY AND CONTROL

cells in awkward corners at the bottom of the combs and tucked in by the sidebars – failure is not an option!

Some books say you must not shake frames or you will dislodge larvae or pupae in the queen cells. However, as long as you shake gently there should not be a problem. If the colony has only recently swarmed (fewer than four days previously), there will be eggs and/or young larvae present from which emergency queen cells can be made. You will need to return in 3–5 days' time and check that none have been started as this could enable a cast swarm to occur, undoing your previous good work.

Don't miss queen cells in awkward corners

A better option for the more experienced beekeeper – one that does not require the destruction of queen cells – can be found in Step 7. We have used this method exclusively over the past 10 years and so far it has been 100% successful. Above all, it avoids the beekeeper choosing the new queen and allows the bees to do it for themselves.

STEP 6 – The colony has definitely swarmed and is left with a reduced number of bees, brood and numerous queen cells

Investigation

This colony will almost certainly produce a cast swarm if you do not do something to prevent this happening. How urgent the matter is depends on how long ago swarming occurred and how mature the queen cells are at the time of your inspection.

If you saw the colony swarm or caught a swarm which you know came from this hive then you are already in possession of the information you require.

RE-ACTIVE SWARM CONTROL
(PRELIMINARY INVESTIGATION AND DIAGNOSTIC TREE)

If you do not have this information you can find out by looking at the brood. Search for the youngest brood (there may be very little of it so look diligently) and count back to day zero. That will tell you roughly how long it is since the swarm took place.

How mature are the queen cells? If you find an open queen cell or queens are starting to emerge as you look through the hive, you are in luck. Go straight to Step 7 because this is a 'golden opportunity'. However, if the cells are not mature you can choose not to follow the method of thinning them down to one. Instead, open a few queen cells, estimate a date when they will be mature and again go straight to Step 7.

There seems to be a two to three day window (a safety margin) after the first virgin has emerged and before the first cast swarm issues. Because the remaining (un-emerged) queens are likely to be more mature at the time of their release, this is probably not the case for subsequent cast swarms.

Remedial Action

This is the same as for Step 5.

STEP 7 – My colony has swarmed and there are emerged and sealed queen cells present

Investigation

This sounds like a rather tricky situation but this is not the case. If the colony has already cast-swarmed, it has happened and there is nothing you can do about that – it is also difficult to be sure unless you have seen or caught the swarm. The best indication is a still more marked reduction in the number of bees in the colony.

If it has not cast-swarmed then you are in luck and it is usually possible to prevent this happening.

SWARMING BIOLOGY AND CONTROL

Open queen cells indicate virgin queens are ready to emerge

Remedial Action

Examine the remaining sealed queen cells which are probably on the point of emerging anyway. You may find that queens start to emerge from their cells as you look through the hive. This happens because your blundering around has distracted the 'warder' bees that were keeping queens penned in until the colony wanted them to emerge. The bees have a plan (which almost certainly includes further swarming) that you are now going to upset!

Investigate the sealed cells carefully, opening them with a knife blade or scalpel and, if the queens are mature and ready to go, help several of them to walk out into the hive – the more the merrier. The point is that you do not know if there is already a virgin queen (or queens) loose in the hive so you are making sure by releasing what are often referred to as 'pulled virgins'. Having had your fun releasing virgin queens into the hive, you now have to do what it says in Step 5 and carefully destroy ALL the remaining queen cells (sealed or unsealed).

Releasing all these queens into the hive at the same time sounds counter-intuitive but seems to force the colony to select from the available talent and settle down to get their chosen queen mated. No matter how many queens are released (and we have released up to 18!) it does not seem to upset the colony and none has ever swarmed after following this procedure. We have done it many times, and so have others, and there have been no failures.

RE-ACTIVE SWARM CONTROL
(PRELIMINARY INVESTIGATION AND DIAGNOSTIC TREE)

STEP 8 – I think the colony has recently produced a cast (secondary) swarm

Investigation

This is a very similar situation to Step 7.

You need to find out if there are any un-emerged queen cells in the hive. You also need to look at any brood to find out how long it has been since the prime swarm departed.

A cast swarm

Remedial Action

If there are still some sealed queen cells, do the same as in Step 7 and release some virgin queens, then destroy ALL remaining queen cells.

At this stage in the swarming process the beekeeper should be very suspicious of queen cells that are still sealed because they are likely to be in this condition because the occupant is DEAD. If there are no sealed queen cells you have to decide whether or not there is a virgin queen in the hive.

There probably is but, if you want to be sure, you can insert a 'test' comb containing eggs and young larvae taken from another hive. If in a few days' time there are no emergency queen cells on this frame, this a clear indication that there is a queen present and it is simply a matter of waiting for her to start to lay.

If queen cells are produced, then there is no queen and you can either allow these cells to produce a new queen or, if you have a source of a queen or sealed queen cells from another colony, you can introduce these to the hive and save time (until there is a new laying queen).

SWARMING BIOLOGY AND CONTROL

STEP 9 – My colony has no unsealed brood, a limited amount of sealed brood and no sealed queen cells

Investigation

The age of the brood will give some idea when the original swarm occurred. Uncap a few worker cells and assess the age of the occupants.

There should be a virgin queen in the hive so you need to ask yourself whether the colony is behaving in a queen-right manner. Does it seem settled or are the bees running around, fanning their wings and making a 'roaring' noise? Another question to ask: are there any laying arcs? These are semicircles of cells on combs, usually in the middle of the hive, that have been kept free of nectar and are highly polished ready for a queen to lay in them.

Frame with no unsealed brood

Neither of these 'signs' is completely foolproof.

Remedial Action

There is not much you can do in this situation except check if the colony has a queen by inserting a 'test' comb and see if queen cells are made (see Step 8). It is always advisable to do the test sooner rather than later because, if there is no queen, you have just been wasting time waiting for something to happen. If the message is in the affirmative (no queen cells are made), all you can do is wait for the queen to start to lay.

STEP 10 – My colony has no brood and no sealed queen cells – help!

Investigation

You now have very little information to tell you what has

RE-ACTIVE SWARM CONTROL
(PRELIMINARY INVESTIGATION AND DIAGNOSTIC TREE)

happened and when it happened. You may be able to see the remains of some queen cells but it will not be possible to tell how old these are. All possibility of swarming is now over and it is just a question as to whether or not this colony will get a new laying queen.

A comb with no brood and no queen cells

Remedial Action

Again you can use the 'test' comb method to determine the queen status as described in Steps 8 and 9 – sooner rather than later.

STEP 11 – My colony has no brood apart from that on a 'test' comb that it may have received previously, but NO queen cells have been produced

Investigation

Failure to make queen cells on the 'test' comb is because the colony thinks (or thought at the time you introduced the frame) that it has a queen – in other words there is (or was) a source of queen pheromone in the hive.

*This 'test' comb has queen cells indicating a queen **is** present*

The first question in this situation is, how long is it since a queen cell could have emerged? There is probably no way to tell by looking at the combs so, if you have no information from previous inspections, you cannot answer this question. If you do know the probable emergence date from your hive records then it is reasonable to expect a laying queen in three weeks or, at the absolute limit, four weeks. Be aware that queens that are late coming into lay are subject to a higher failure rate (at some time in the future) than queens which start to lay on time.

You can assess the behaviour of the colony as described in Step 9; is it behaving in a calm manner, does it have laying arcs? You can also look to see if you can see a queen but non-laying queens are difficult to find. You could give the colony another 'test' comb but time is running out for them to be successful in producing a new queen.

Remedial Action

In this situation, to be successful in re-queening by any method (using a frame with eggs and young larvae, a sealed queen cell or a queen) you must find the source of queen pheromone and eliminate it. You need to search diligently for a queen and, if you find her, you must kill her – providing you are convinced she is never going to lay. If you have been successful in this task you will now be able to re-queen the colony. However, introducing a mature queen cell from another colony or a laying queen is likely to be a better option than letting the colony raise its own queen at this late stage.

Step 12 – My colony has got a drone-laying queen

Investigation

A drone-laying queen can be identified by the presence of brood in worker cells that has domed cappings – like the capping on a drone cell but on a smaller worker cell. This means that the queen is laying unfertilised eggs when she should be laying fertilised ones.

This can be from a variety of causes: she may not have mated properly, she may have run out of sperm, she may have some

RE-ACTIVE SWARM CONTROL
(PRELIMINARY INVESTIGATION AND DIAGNOSTIC TREE)

internal defect or she may have a deformed wing virus infection. Initially, such a queen may lay both fertilised and unfertilised eggs, producing a mixture of normal and abnormal brood in worker cells.

Things can only get worse (not better) so now is the time to take action.

Brood of a drone-laying queen

Remedial Action

This is exactly the same as for Step 11 – you must find the queen and eliminate her before you can re-queen.

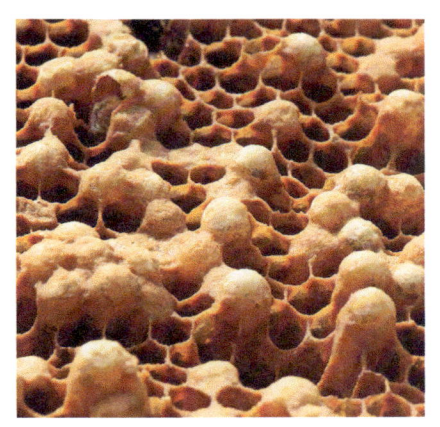

Laying worker brood

Occasionally, worker brood with domed cappings is not produced by a drone-laying queen but by laying workers. This is not easy to diagnose. The signature of laying workers is that the brood is patchy with little regular pattern, eggs may be laid on the side walls of the cells and there may be more than one egg per cell.

It is said that a colony with laying workers will not accept a mature queen cell or even a laying queen but usually there is no problem. If the colony is still worth saving (or rather the bees in it are worth saving) the safe option is to unite it with a queen-right colony.

Postscript

If you have followed the steps in the above key diligently, at each stage making a careful study of the evidence available in the hive, you should now be in a position to manage that colony with the best chance of a favourable outcome – for both the bees and the beekeeper. You will not always be successful, either because you have failed to arrive at the correct diagnosis (it is not always easy) or because of matters outside of your control, such as queen mating, but you will have given it your best shot. Honey bee colonies are all different and this is part of the challenge and fascination of beekeeping.

2.4 RE-ACTIVE SWARM CONTROL (ARTIFICIAL SWARMING AND ASSOCIATED MANAGEMENT)

Re-active swarm control starts when pre-emptive swarm control fails – when queen cells with contents are found in the hive. As previously emphasised, when this happens there is no known (reliable) method by which the swarming process can be turned off and the beekeeper needs to accept the fact that unless something is done to prevent it, that colony will inevitably swarm.

Destroying queen cells only delays swarming and may make the situation worse, ie, the colony swarms with barely started second-generation queen cells and takes a higher proportion of the bees with it. Worse still is destroying queen cells not realising that the colony has already swarmed. If swarming occurred more than four or five days earlier there will be no eggs or young larvae present from which emergency queen cells can be made and the colony will be rendered queen-less.

Diagram conventions

These are the same as those set out in Chapter 2.3. Except when otherwise stated, all diagrams show a standard hive configuration of a brood-and-a-half, ie, one deep and one shallow box beneath the queen excluder (the brood area), which seems to be about the right amount of space for near-native (locally adapted) bees in Wales. The number of supers shown is variable according to the likely development or needs of the colony under consideration.

Other hive configurations (eg, single deep, double deep or 14x12) require appropriate modification of the methods described below and some options may be limited or even impossible.

Using a queen excluder

It is also assumed that beekeepers are using queen excluders on their hives during the summer months and, where appropriate, one is shown in the diagrams.

As far as swarm control is concerned, the use of a queen excluder has it pros and cons. I will not attempt to deal with these in detail, but not using a queen excluder potentially gives the queen more space in which to lay and is therefore a form of pre-emptive swarm control. Using a queen excluder makes some manipulations in re-active swarm control easier (because it limits where the queen can be in the hive). But not using a queen excluder certainly does not preclude following any of the management techniques involved but they may have to be slightly modified.

Determining the stage in the swarming process

On finding queen cells in a colony, the first thing to do is to determine what stage in the swarming process the colony has reached. However, before jumping to conclusions, the beekeeper needs to be sure that the queen cells are actually

RE-ACTIVE SWARM CONTROL
(ARTIFICIAL SWARMING AND ASSOCIATED MANAGEMENT)

swarm cells and not supersedure or emergency queen cells. This is usually obvious by their position and number but, if in doubt, refer back to Chapter 2.3 which contains a 12-step diagnostic tree, each step including a method of investigation and the recommended remedial action.

For practical purposes the 12 stages can effectively be reduced to four main stages in the swarming process, each with a different solution:

1. the colony has queen cells but has not yet swarmed
2. the colony has issued the prime swarm but has not yet cast-swarmed
3. the colony has emerged queen cells and may (or may not) have issued a cast swarm
4. there is evidence that the colony has swarmed but it currently appears to be queen-less (ie, it has no brood of any age), and the beekeeper has no idea what happened and when.

The beekeeper should do ABSOLUTELY NOTHING until a careful investigation has been completed. Fortunately the key information persists in the hive (on the combs) until at least three weeks after the prime swarm has departed – the last time the queen can have laid some eggs.

2.4.1 The colony has not yet swarmed

How do you know? The evidence is, in order of reliability:

- the queen is seen
- there are recently laid eggs (standing on their ends)
- there are eggs leaning over at about 45°, which were laid 24–48 hours previously, and those lying flat, laid over 48 hours ago and about to hatch (be aware that a late afternoon inspection may reveal recently laid eggs but a swarm could have departed earlier in the day)

- there are no missing bees (the colony is the size you expect)*
- the maturity of queen cells and recent weather conditions – has it been conducive to swarming?

NB. That there are no missing bees is to be expected if the queen has had her wings clipped – the colony can have swarmed and subsequently returned to the hive and there will be little or no loss of bees. This can also occur naturally if for some reason the queen is unable to fly – be aware of this as it is not exactly a rare occurrence (for further discussion see Appendix 4).

The beekeeper really needs to see the queen or recently laid eggs early in the day to be absolutely sure. If the colony is deemed not to have swarmed, the remedy is to make an artificial swarm.

Artificial swarming

Having established that the colony has not yet swarmed the (only) solution is artificial swarming.

Most beekeeping books describe what is called the Pagden method, first published in 1868. [Originally this described how to deal with the parent colony to prevent cast swarming once the prime swarm had issued, been caught and hived in the original position, the parent colony having been moved to another position nearby. The term is now used to describe a process that is designed to prevent the colony from issuing a prime swarm. Such processes collectively come under the heading of artificial swarming.]

The unreliability of the Pagden method and the underlying causes were discussed at some length in Chapter 1.1, as was the development of a method called Snelgrove II (modified). In many ways the latter is a simpler method for the beekeeper to apply and its reliability has been confirmed over 15 years of use

RE-ACTIVE SWARM CONTROL
(ARTIFICIAL SWARMING AND ASSOCIATED MANAGEMENT)

in our own apiaries. Many other beekeepers have subsequently used the method to good effect.

As the name suggests, the basic method originates with LE Snelgrove, so his name has been retained out of respect for a very distinguished beekeeper (the father of swarm control in my view). The modified version is subtly different from the method as originally conceived by providing the (queen-less) artificial swarm with the means of making emergency queen cells.

This is thought to be the secret of its success by changing the behavioural programme on which the colony is operating from swarming to emergency re-queening, thus effectively switching off the swarming impulse. Despite the name, it does not require the use of a Snelgrove board (or even a split board). It can also be done with the parent colony being moved to a separate hive stand but the use of a split board does have many advantages (see the discussion below).

Artificial swarming by the Snelgrove II (modified) method is a two-stage operation which is illustrated in Figures 11a and 11b using a new hive stand (not a split board in sight!). For clarity, the brood area of the parent colony is in blue boxes (with a blue background) and the artificial swarm is in a green box (with a green background).

The first manipulation

In the initial manipulation (Figure 11a), all the brood, including the queen and queen cells, is moved on its existing floor to a new hive stand within the apiary – reasonably close is convenient but more than three feet away is essential. This is now referred to as the 'parent colony'.

A new floor is placed on the vacated hive stand and a new box containing 10 frames of, preferably, drawn comb (but a mixture of drawn comb and foundation will suffice) is placed on it. This will become the 'artificial swarm'.

SWARMING BIOLOGY AND CONTROL

Parent colony – before

Colony on brood-and-a-half with 10 and 8 frames of brood

Parent colony and artificial swarm – after

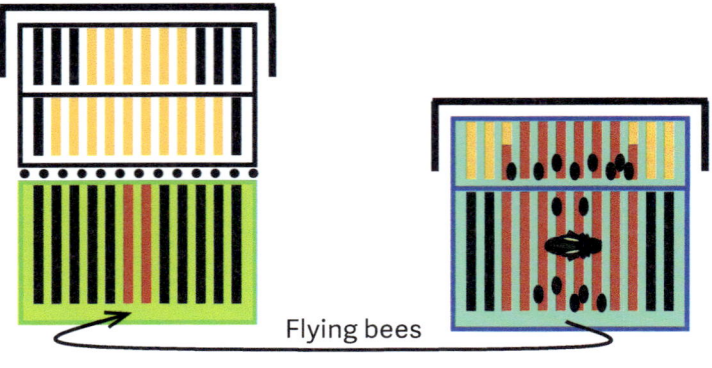

Flying bees

A new deep box on the old stand + two frames of brood to make emergency queen cells

All brood + queen cells + queen on a new hive stand

Queen | Brood | Stores | Empty frame | Queen excluder | Queen cell
Parent colony | Artificial swarm

Figure 11a Snelgrove II (modified) using a separate hive stand – the initial manipulation

RE-ACTIVE SWARM CONTROL
(ARTIFICIAL SWARMING AND ASSOCIATED MANAGEMENT)

Emergency queen cells produced are now destroyed and the queen returned

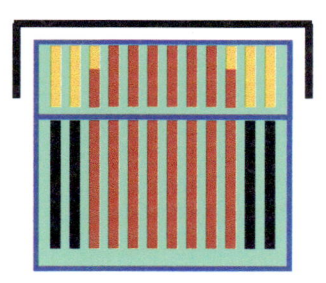

Swarm cells torn down by the bees and the queen has resumed laying

• •

After a further 7–9 days

Emergency queen cells are produced from recently laid eggs and newly hatched larvae

Queen | Brood | Stores | Empty frame | Queen excluder | Queen cell

Parent colony | Artificial swarm

Figure 11b Snelgrove II (modified) using a separate hive stand – the second manipulation after 9–10 days

Two frames of brood are now removed from the parent colony and transferred to the middle of the artificial swarm. These two frames must contain eggs and young larvae from which emergency queen cells can be made but must not have any queen cells (or the queen) on them. If the beekeeper is not confident about spotting the queen, these two frames can have the bees on them shaken off into the parent colony.

The missing frames in the parent colony should be replaced using drawn comb if possible (but dummy boards can also be used).

Both hives are now rebuilt and most (or all) of the supers are normally given to the artificial swarm where the majority of the current foraging force resides. However, remembering that the parent colony is initially going to lose all its flying bees, the quantity of available stores needs to be taken into consideration. If these are thought to be insufficient to last until a new foraging force has developed or there is unlikely to be a significant nectar flow in the near future, the parent colony should be given a queen excluder and one (or possibly more) of the supers. In some seasons it may even have to be fed.

Losing the flying bees

The first thing that happens is that the parent colony on the adjacent stand will lose its flying bees back to the artificial swarm. Amongst these will be the bees that are running the swarming process (the swarm organisers). For this reason, the parent colony completely loses the impulse to swarm and, in due course, the queen cells will be torn down and the queen will resume laying.

Emergency queen cells

In the queen-less artificial swarm, the bees will start emergency queen cells – and this appears to be vital to the loss of the swarming impulse.

RE-ACTIVE SWARM CONTROL
(ARTIFICIAL SWARMING AND ASSOCIATED MANAGEMENT)

The second manipulation

The second manipulation should take place 9–10 days later (12 days is the absolute maximum safe limit).

Timing is important because all the queen cells in the parent colony should be torn down by this time. Usually this happens quite quickly but it does depend on the maturity of the queen cells, ie, it is not done until the virgin queens they contain are 3–4 days from emergence at which time the colony becomes fully aware of their presence. However, if the queen cells were only just started at the time of the split, it may take a day or two longer.

More importantly, the second manipulation must be done before any virgin queens can emerge into the artificial swarm.

The calculation goes like this: if an emergency queen cell is based on a one-day-old larva (Day 4 from when the egg was laid), the first queen could emerge sometime on Day 12. From this it can be seen that Day 12 is pushing the limit and it is better not to take the risk.

Check for the presence of the queen

The second manipulation starts with checking that the queen cells in the parent colony have been torn down and the queen has resumed laying. If this has not happened it means that the beekeeper has somehow failed to transfer the queen to the parent colony.

The most likely cause for this is that the beekeeper failed to notice that the colony had already swarmed (sometimes it is difficult to be absolutely sure). This is not a cause for alarm because the parent colony has queen cells from which it can re-queen itself (or by this time may already contain a virgin queen) and the artificial swarm has emergency queen cells approaching maturity. The parent colony will re-queen without swarming (no management required) but steps may be required to prevent cast swarming by the artificial swarm (see Section 2.4.2).

There are other less likely explanations for the queen cells in the parent colony not having been torn down and the beekeeper should refer to Appendix 3 to understand this situation and what to do next.

The next step is to examine the artificial swarm where the two frames which contained eggs and young brood should now have a number of emergency queen cells on them. If this is not the case, the queen (or a queen) must be present somewhere in the artificial swarm – again refer to Appendix 3.

Transfer the queen

The frames with queen cells should now be removed and then the queen can be transferred to the artificial swarm. Details of the best way to do this (smoothly and safely) can be found in Appendix 2.

The emergency queen cells

The question now is what to do with the two frames with emergency queen cells on them?

If at the time the colony was artificially swarmed there were plenty of younger bees in the supers (and this is usually the case with a colony that has set up to swarm), then the occupants of the emergency queen cells will have been well-fed and contain fully developed virgin queens. If the beekeeper is confident this is the case, the two frames with emergency queen cells can be placed in the (now queen-less) parent colony to provide it with a new queen.

If there is any doubt about the quality of the cells then it is better to destroy them and let the parent colony start from scratch and make its own. It will have available suitable brood produced by the queen which had resumed laying during her sojourn in the parent colony. Also there will be plenty of nurse bees from the still-emerging brood to feed them well. The latter is the safer option but the former saves 9–10 days on the re-

RE-ACTIVE SWARM CONTROL
(ARTIFICIAL SWARMING AND ASSOCIATED MANAGEMENT)

queening of the parent colony. Which option to take requires a judgement call to be made by the beekeeper.

Artificial swarming using a split board

Starting with an identical colony that has set up to swarm, Figures 12a and 12b show the equivalent process using a spilt board (the colour coding of the boxes is the same as in Figures 11a and 11b). The overall process is just the same but transfer of the queen to the artificial swarm in the second manipulation requires slightly different logistics, which are described in Appendix 2.

If in the first few days after the first manipulation it is noted that the bees in the artificial swarm have found the queen (see page 14), the only remedy is to immediately move the parent colony to a new position in the apiary so that contact is broken. This is a fairly remote possibility but see the following two paragraphs for precautions that can be taken.

Looking for the queen

When using a split board, the first reaction of an artificial swarm on losing its queen is to frantically look for her. If she cannot be found they will then start emergency queen cells. However, if the vertical separation between the parent colony and the artificial swarm is small, eg, one super, then there is a small chance that they will succeed in finding her.

I suspect this happens because the bees that do the finding are not those that are directly involved in making the queen cells and it takes time for them to recognise the current situation. Two supers separation is much safer and three is completely secure. Although the beekeeper needs to be aware of this problem and its implication, colonies that have received good pre-emptive swarm control will almost inevitably have several supers on them when they finally set up to swarm (so the problem does not arise). If this is not the case, it would

SWARMING BIOLOGY AND CONTROL

Parent colony →

Artificial swarm　　**Flying bees**

The original colony on brood-and-a-half + two supers; set up to swarm

The complete brood nest along with the queen cells and the queen is moved to the top of the hive on a split board BUT two frames of brood, without queen cells, are transferred to the artificial swarm

Queen　Brood　Stores　Empty frame　....Queen excluder　Queen cell

Parent colony　Artificial swarm

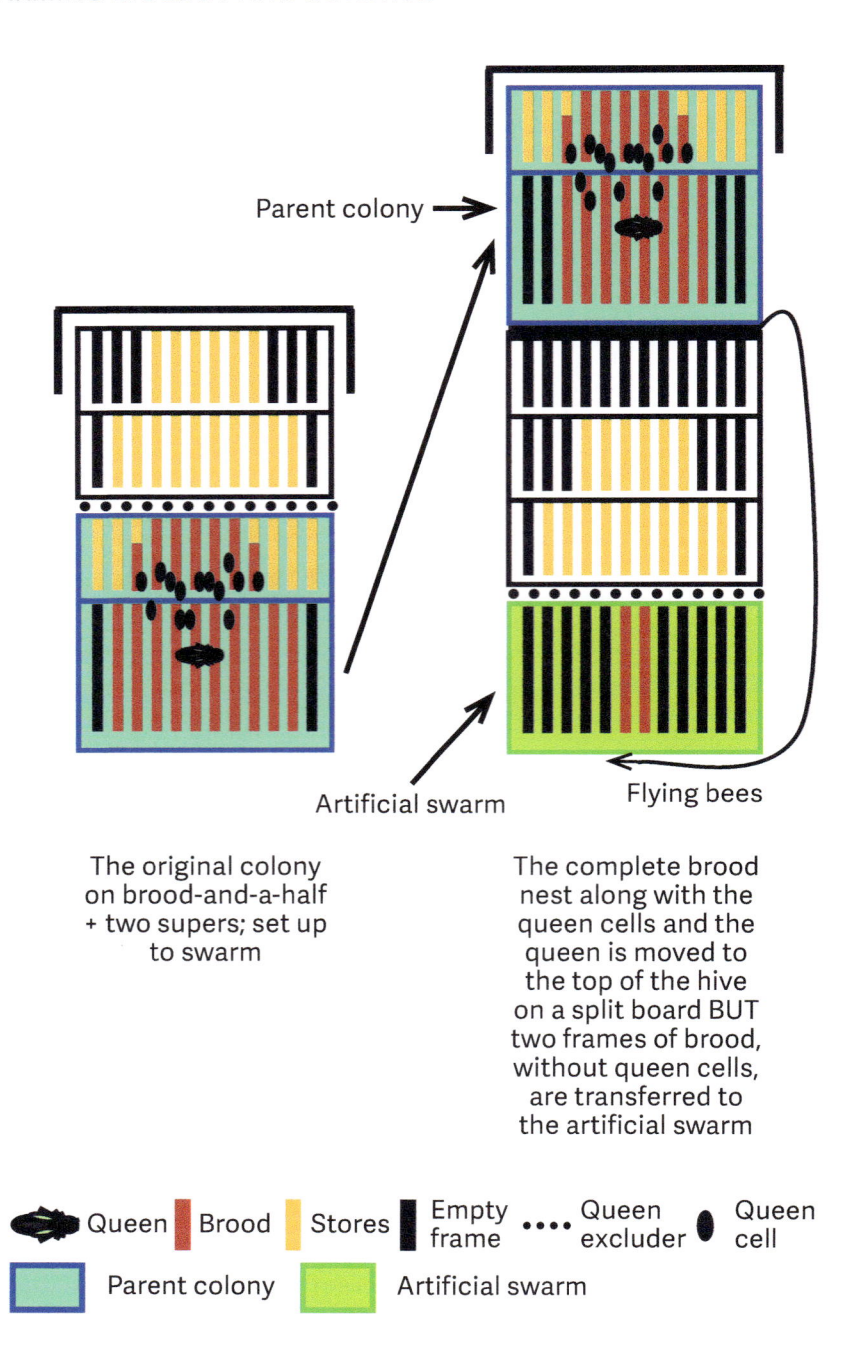

Figure 12a Snelgrove II (modified) using a split board – the initial manipulation

RE-ACTIVE SWARM CONTROL
(ARTIFICIAL SWARMING AND ASSOCIATED MANAGEMENT)

Queen cells at the top of the hive have been torn down and the queen has resumed laying. There are emergency queen cells in the artificial swarm

The queen has been returned to the bottom of the hive and the emergency queen cells destroyed. New emergency queen cells will be started in the top colony

Legend: Queen | Brood | Stores | Empty frame | Queen excluder | Queen cell | Parent colony | Artificial swarm

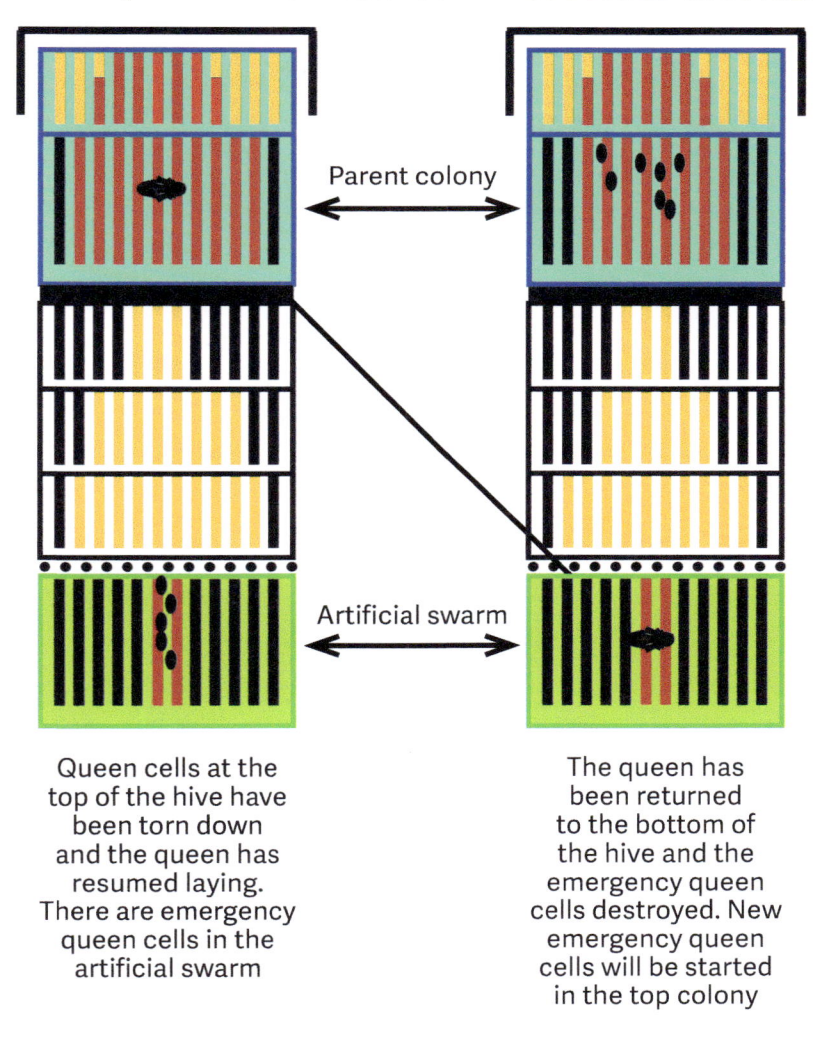

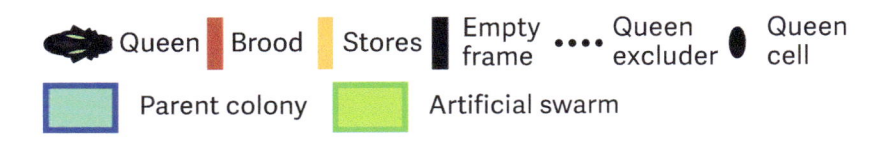

Figure 12b Snelgrove II (modified) using a split board – the second manipulation 9–10 days later

be prudent to add an extra super when making the artificial swarm just to give greater vertical separation, even though it is not required for the storage of honey.

The alternative is to use the two-stand version (Figures 11a and 11b), where bees from the artificial swarm (more than three feet away) will be unable to find where the queen has gone.

Advantages of Snelgrove II (modified)

Apart from its virtually 100% success rate, one of the main advantages of this method of artificial swarming is that the queen does not have to be found to undertake the first manipulation. Even for an experienced beekeeper, finding the queen is often not easy when the colony is crammed with bees and the queen is being constantly harassed by the workers to slim her down. She could be anywhere and is probably running as fast as she can to get away from her assailants.

When she does have to be found in the second manipulation, there will be fewer bees and she will have settled back to the day job (laying eggs) – so finding her should be much easier.

Got the timing wrong?

If, for some reason, the beekeeper gets the time of the second manipulation wrong (eg, because of other commitments or the weather was awful) and the emergency queen cells have already released virgins into the artificial swarm, it is not the end of the world. It is unlikely that all the virgin queens can be found and removed and the old queen safely transferred to the artificial swarm so it is better to allow the artificial swarm to re-queen naturally and leave the old queen where she is in the parent colony. This part can now be given a queen excluder and some supers and will become the main honey-producing unit.

However, there is conflicting testimony from users as to what happens to the artificial swarm which still contains multiple

RE-ACTIVE SWARM CONTROL
(ARTIFICIAL SWARMING AND ASSOCIATED MANAGEMENT)

queen cells. Does it behave in the same way as a normal colony that is following an emergency re-queening programme and select a new queen without swarming? Or, finding itself with multiple queens, does it proceed to issue a swarm (or swarms)? Or are both outcomes possible?

At the present time this is not known and a safe way of dealing with this uncertainty is given in Section 2.4.3.

The aftermath of artificial swarming

If successful (ie, the new queen has mated and started laying), when the process of artificial swarming is complete there will be two colonies instead of one and this may or may not be what the beekeeper wants.

If an increase in the number of colonies is not required then uniting is the answer. This is easier when using a split board than it is with the new colony on a separate hive stand. However, uniting should not (normally) be attempted until the parent colony has developed a new laying queen and she has proven herself to be a good 'un.

A two-tier hive

Using a split board, a two-tier hive (one colony on top of the other, each with its own supers) can be retained until the end of the season when a decision as to its future can be made.

Note: this is not a two-queen system because the two colonies are independent and have different entrances (facing in different directions) and separate supers. However, be warned that in a good season a two-tier colony can become a tower block and, unless you are both tall and strong, two-person handling is advisable (or essential – I joke not!).

Hive inspection will become difficult because of all the heavy lifting required. Fortunately a colony that has been artificially swarmed by this method rarely attempts to swarm again that

A two-tier hive often requires handling by two people

season so routine inspection, checking for the start of queen cells, can be relaxed without incurring serious risk. Providing both colonies have adequate supering, there should be nothing requiring intervention by the beekeeper.

The main problem comes at the end of the season, when it is time to harvest the (hopefully heavy) honey crop. Using clearer boards, this usually has to be done in two sessions – the top colony first followed by the bottom one, usually on a different day. This is a good time to decide on the future of the two colonies.

RE-ACTIVE SWARM CONTROL
(ARTIFICIAL SWARMING AND ASSOCIATED MANAGEMENT)

Uniting colonies

When uniting two colonies it is preferable for the beekeeper to choose the queen because letting the colonies do it for themselves can result in serious conflict. Also, I find it difficult to believe the old adage that 'the bees will always choose the best queen'. From the beekeeper's point of view, the new queen is usually preferable.

If extra colonies are required then, with the new colony on a separate stand, this has already been accomplished. If a split board has been used, then moving the new colony to a different apiary to give it independence is preferable as there will be no loss of flying bees. If a new colony on a split board is to be kept in the same apiary, care must be taken to avoid excessive loss of flying bees. It is recommended that 'independence day' should be delayed for a minimum of five to six weeks after the new queen has started to lay – which, in reality, may be pretty much the end of the season.

2.4.2 The colony has already issued the prime swarm but has not cast-swarmed

It is assumed that the beekeeper has already determined that the prime swarm has departed with the old queen. It is then just a matter of carefully checking that no queen cells are open, releasing a virgin queen into the colony. The age of the youngest brood gives some clues, eg, if there are no eggs but some newly hatched larvae then swarming occurred 3–4 days ago and, under normal circumstances, the queen cells are about four days from emergence.

In this situation, unless something is done to prevent it, the colony is likely to issue at least one cast swarm. The timing of this swarm will be 2–4 days after the first virgin has emerged. There are two methods of preventing this (previously discussed in Chapter 2.3). The alternatives are:

- to reduce the number of queen cells to one (see Chapter 2.3, Step 5)

SWARMING BIOLOGY AND CONTROL

- wait until the queen cells are mature and virgin queens are emerging and release some more, followed by destroying all other queen cells (see Chapter 2.3, Step 7).

2.4.3 The colony contains both emerged and sealed queen cells and may (or may not) have issued a cast swarm

There is no easy way of knowing whether the colony has already cast-swarmed and no way of knowing for sure if the colony contains a virgin queen (or queens). Most of the remaining queen cells will contain queens that are waiting to emerge but are being prevented from doing so by 'warder' bees posted on the cells. Using the tip of a knife or scalpel, the beekeeper should carefully open several cells and let virgin queens walk out into the colony (see Chapter 2.3, Step 7).

2.4.4 The colony appears to be queen-less (it has no brood) and the beekeeper has no idea what happened and when

There may be a queen in the colony which is just about to start laying but, without actually finding her, you cannot tell. The behaviour of the colony (the bees seem calm) and the presence of laying arcs (cells prepared for a queen to lay in) imply that all is well with the colony but neither of these signs is completely reliable.

The best thing to do in this situation is to insert a 'test' comb (taken from another colony) containing eggs and young larvae. If the colony is queen-less, emergency queen cells will be made on this frame but if the colony considers it has a queen, the donor brood will be raised in the normal way. Even this is not 100% reliable because in rare cases the colony may contain a non-laying queen (which is producing enough queen substance to make the colony think they do have a queen) and no further progress can be made (ie, the colony cannot be re-queened) until she has been found and removed (see also Chapter 2.3, Step 8).

RE-ACTIVE SWARM CONTROL
(ARTIFICIAL SWARMING AND ASSOCIATED MANAGEMENT)

Late-season swarming

As already noted, this can be one of the unfortunate by-products of pre-emptive swarm control.

Typically, a large colony that has been kept together with no attempt at swarming until late-June or early-July will suddenly develop queen cells. This could still be controlled by means of artificial swarming (as described above) but splitting the colony just when the main nectar flow is about to start is definitely the last thing the beekeeper wants.

One method of dealing with this situation is to simply remove the queen. You do not need to do anything drastic like killing her (you will probably want to repatriate her later), you merely need to put her aside (in 'purdah') where she can safely tick over because her services may be required later (see below). It is not a perfect solution but it works well enough.

As soon a queen cells are seen, the queen should be located and transferred (on the frame on which she has been found) to a nucleus box. To this should be added 2–3 frames of brood and the box made up with frames of stores. Additional support for the queen should be added by shaking in extra bees from several combs. This box should now be moved to a new location somewhere in the same apiary.

Another way of putting the queen in purdah is to do it over a split board on top of the hive (Figure 13).

As soon as queen cells are seen the queen is removed along with 2–4 frames of brood and placed in a new box (preferably a shallow one if using a brood-and-a-half configuration) which is placed on a split board at the top of the hive. This box should be made up with drawn frames (some of them containing stores) and, finally, additional support for the queen should be given by shaking in a generous number of extra bees.

Don't forget that this is a time of year when robbing may occur so, using either method, there needs to be enough bees and a small entrance to prevent this happening.

SWARMING BIOLOGY AND CONTROL

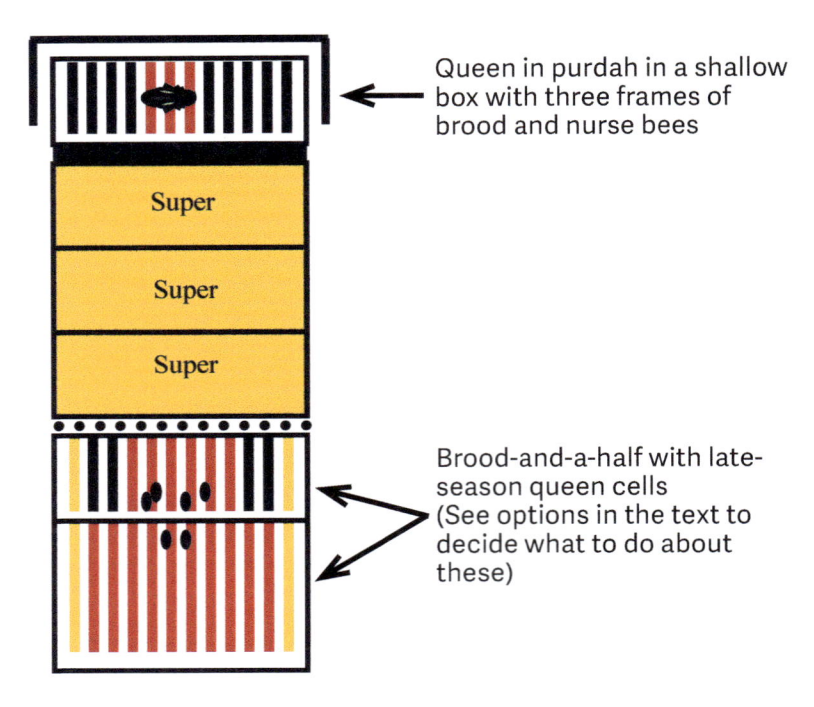

Figure 13 — Diagram showing a hive configuration:
- Top: Queen in purdah in a shallow box with three frames of brood and nurse bees
- Middle: Three Supers
- Bottom: Brood-and-a-half with late-season queen cells (See options in the text to decide what to do about these)

Figure 13   Late-season method of dealing with a colony set up to swarm by putting the queen in purdah

What to do with the queen cells?

The best option is probably to destroy the queen cells but only after ensuring that there are eggs and young larvae present from which the colony can start emergency queen cells. The colony will probably be switched into emergency re-queening mode and when these cells are mature (about 12 days later) the bees will choose a new queen and (hopefully) not swarm.

If you want to be sure that swarming does not occur then follow the method of releasing virgin queens as described in Chapter 2.3, Step 7.

NB. When using the 'purdah' method, be aware that queens mated later in the season (from July onwards) are known to be less reliable (less likely to survive their first winter) than those mated earlier (May and June). It is thought this is due to

RE-ACTIVE SWARM CONTROL
(ARTIFICIAL SWARMING AND ASSOCIATED MANAGEMENT)

the increased probability of the queen mating with a drone(s) that is carrying deformed wing virus (DWV) and becoming infected herself – in other words a sexually transmitted disease (STD). For this reason, 9–10 days later it is better to destroy the emergency queen cells and repatriate the old queen, as in the second manipulation for the Snelgrove II (modified) method described above.

During this enforced queen-less period, the colony will continue to forage well (but possibly not quite as well as it would with an incumbent laying queen) and there will be little loss of honey crop.

A new method of dealing with a late-swarming colony

We have been experimenting with methods of dealing with late swarming and so far (over the past three years with more than 20 replications), the following method has proved reliable.

The manipulation involved is shown in Figure 14 and, perhaps unwisely, we have called it the Demaree method. However, please do not confuse this with the true Demaree which, as a method of pre-emptive swarm control, has totally different aims. It does have a superficial similarity in that it has a box containing brood at the top of the hive but this is nothing to do with making room for the queen to lay or attracting nurse bees from the main brood area.

The colony has been artificially swarmed and, as shown in Figure 14, is in the same condition as it would be for the second manipulation of Snelgrove II (modified) – the queen cells in the parent colony have been torn down, the queen has resumed laying and there are emergency queen cells in the artificial swarm.

Initially the manipulation proceeds the same way; the emergency queen cells are removed from the artificial swarm and (unless there is some use for them elsewhere) are destroyed and the queen is transferred to the artificial swarm.

SWARMING BIOLOGY AND CONTROL

Queen cells torn down; queen has resumed laying

Snelgrove board

Queen transferred to the artificial swarm

All emergency queen cells must be destroyed

The colony 9–10 days after the artificial swarm

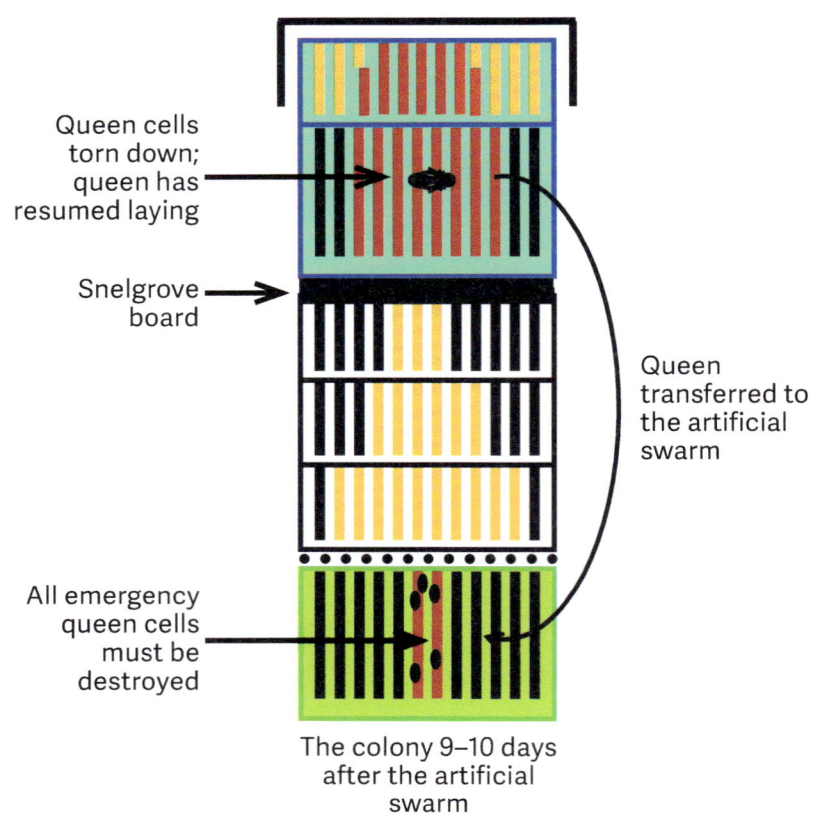

Figure 14   Demaree method of reuniting a late-swarming colony

Now the mesh from the middle of the split board is removed so the bees in the parent colony and artificial swarm can mingle – which they should do without any fighting.

The aim now is to remove the split board as soon as possible but before this can be done the bees in the parent colony need to be re-trained to use the entrance at the bottom of the hive instead of that on the split board to which they are accustomed (see the explanation below). Bees are much better at adapting to a change in the elevation of the entrance than they are to a change in orientation (the direction from which they approach the entrance). So the flying bees on the split board first need to be redirected horizontally.

RE-ACTIVE SWARM CONTROL
(ARTIFICIAL SWARMING AND ASSOCIATED MANAGEMENT)

When using a Snelgrove board, this can be easily (and more gently) achieved by opening and closing the appropriate doors. Leaving the door you are persuading them not to use slightly open (one bee width) and having the door you want them to use (the one at the front of the hive) wide open will speed the adjustment process.

With a single-entrance split board it can only be done by turning the board itself and, if the change in direction is only 90°, the bees will quickly adjust to this. However, a change of 180° needs to be done in two stages but, in good weather the bees only need an hour or so to adjust, so the direction of the entrance can be changed twice on the same day.

The explanation

It is essential that the bees in the box containing brood at the top of the hive (which was the parent colony) are forced to descend through the hive to the bottom, where the queen now resides, in order to fly. This is to ensure that they come into contact with enough queen substance to realise that they are queen-right, otherwise they will start emergency queen cells at the top of the hive.

The desired result has now been achieved and, after a limited separation (9–10 days), the whole colony (and particularly the foraging force) is back together as one unit for what remains of the season. In due course the brood at the top of the hive will emerge and the bees will move down the hive to wherever their services are needed.

Depending on how long nectar continues to flow, honey may be stored in the vacated combs and can either become part of the honey harvest or the frames can be moved down to the bottom of the hive to serve as winter stores.

Conclusions

The combination of pre-emptive and re-active swarm control (Chapters 2.2–2.4) should provide the beekeeper with a comprehensive package of management techniques by which the swarming impulse of honey bee colonies can be kept under some measure of control. The qualification 'some measure' is used because the beekeeper who claims to have achieved complete control is probably 'not of this world' or is being economical with the truth.

Honey bee colonies have individual characteristics and the most common type to evade swarm control are those that start queen cells and swarm long before the first one is sealed, thereby avoiding detection by routine inspection (a so-called Houdini colony). Even with a 6–7-day inspection interval during the swarming season it is impossible to prevent this sort of thing from happening occasionally.

To bring the subject of swarm control full circle, I return to what I said at the end of the second paragraph of the Introduction to Part 1: *There is no doubt that swarm control is simultaneously the most important and most difficult aspect of colony management with which the beekeeper has to deal – if they chose to do so, of course.*

However, if you have read the book to this point, I assume you do want to practice some degree of swarm control.

APPENDICES

Appendix 1
It is not clear whether the queen cells are for swarming or supersedure

If the queen cells are sealed or near sealing and a swarm could be imminent then the safest option is to split the colony. The queen should be left where she is and the queen cells, along with several frames of brood (3–4), and sufficient bees to look after them, should be put in a nucleus box or in a box over a split board at the top of the hive. The queen cells in the split will mature and in due course a virgin queen will emerge. Because all the flying bees have returned to their original site (where the queen is), no swarming will occur. If left in this condition, the new queen should be able to mate and start to lay.

It is subsequent events in the main colony that need to be followed closely.

If a new batch of queen cells is started immediately (and there may be more of them this time) then it is almost certain that the original intention was swarming and the colony will need to be artificially swarmed.

If no more queen cells are produced then it was probably an attempt at supersedure. In this case, the performance of the original (now suspect) queen needs to be monitored: is she producing a satisfactory amount of brood? Further supersedure cells may or may not be produced in the short term.

If the original queen continues to perform well, then nothing needs to be done until the new queen is up and running and then the decision made whether (or not) she should be used to replace her mother as the colony originally might have intended.

Appendix 2
Recommended logistics for Snelgrove II (modified), the second manipulation

If the queen cells in the parent colony have been torn down and the queen has resumed laying there will automatically be emergency queen cells in the artificial swarm. This is a sign that the first manipulation has gone according to plan.

At Day 9–10, the emergency queen cells will have reached a stage in their development when the bees will have started to take an interest in them and will realise that they are soon to be saved by the emergence of a new (virgin) queen. So when the queen cells are removed and the bees suddenly find a fully mated queen in their midst (a miracle has occurred!) they seem to take this in their stride and accept her without question – at least this is the explanation behind what happens (we think).

It is always best to return the queen on the frame where she is found, surrounded by her own nest mates. She can, however, be introduced on her own, and we have done this a few times with no problems. Other people have expressed concern about this practice (even reported that the queen had been killed). So, to avoid that possibility, this is a safer way to ensure her acceptance by bees from which she has been separated for 9–10 days. Using a split board is inherently safer because the common hive odour is retained.

The first task in the second manipulation is to find the queen and this is best done before there has been too much disturbance. In order to do the manipulation smoothly, the frame she is on should be removed and placed carefully in a holding box where she will be safe. A nucleus box (with its entrance blocked) can be used for this purpose but we have a purpose-built box that holds a single frame (firmly held in a castellation at each end). This box comes in useful for other manipulations where you want to ensure the safety of the queen while you are messing about in the hive.

APPENDICES

A single-frame box for holding the queen

The next job is to remove the frame (or frames) with the emergency queen cells on them from the artificial swarm. The frame with the queen on can now be carefully inserted into the space thus created. If two frames were removed, a second frame from the parent colony (complete with bees on it) can be used to fill the space.

When the artificial swarm has been made using a split board, the logistics of this manipulation are slightly more difficult because the parent colony has to be removed (on its board) and set on one side to gain access to the artificial swarm. It helps to have a spare box or roof available on which the parent colony can be temporarily placed.

It is then just a matter of deciding what to do with the emergency queen cells; whether to destroy them and let the parent colony re-queen itself from scratch or transfer them intact to the parent colony to provide a new queen 9–10 days earlier (see the discussion on page 184).

The last task is to reassemble the colonies, complete with supers, adding extra supers if required. Bear in mind that the

SWARMING BIOLOGY AND CONTROL

colony on the split board has yet to re-queen and this should be disturbed as little as possible until there is evidence that the new queen has started to lay.

However, if access to the bottom part of the hive is required for some reason, it should be done outside potential mating hours, ie, before around 10 am or after 5 pm.

Appendix 3
Fault finding

A few people who have used the Snelgrove II (modified) method have reported that it has not worked for them. With one exception (details below) they have been unable to provide sufficient details to determine the reason for the failure.

Over the years we have had one failure in well over 150 replications but are confident we understand the cause (details below). However, there are other possibilities* and the following is a fault-finding checklist.

NB. These are just logical possibilities and we do not know for sure whether anybody has ever made any of these mistakes.

1. As noted in the main text, the second manipulation (9–10 days later) starts with checking that the queen cells in the parent colony have been torn down and the queen has resumed laying. If the queen cells have not been torn down and there are no eggs to be found then a mistake was made in the initial manipulation, ie, the queen is not in the parent colony and she has either been lost altogether or is still somewhere in the artificial swarm.

2. The next step is to check the artificial swarm to see if emergency queen cells have been produced on the two donor frames. If there are queen cells then the missing queen cannot be in the artificial swarm. As already noted, the most likely cause is that the colony had already swarmed but there are other possibilities – the queen had been killed, dropped on the ground or flown off.

3   If there are no queen cells then the queen is still residing somewhere in the artificial swarm. Another possibility is that the frames did not contain brood young enough for the production of emergency queen cells (but this would have been a very silly mistake to have made!).

4   If there are eggs and young larvae in the brood box containing the artificial swarm then you have inadvertently created a Pagden-type artificial swarm, in which case it will probably either have worked or not worked by now (9–10 days later).
   - If it has swarmed then there should be no queen present and the age of the youngest brood will tell you when this happened.
   - If it has not swarmed (but still could do so) there will be a queen and recently laid eggs but you will need to check for the start of queen cells for about another 10 days.
   - In both cases, to prevent further swarming you need to thin the number of queen cells or release virgin queens as recommended in Section 2.4.3 in the main text.

5   If there are no queen cells in the artificial swarm there is one final possibility and this is that the queen somehow found her way into the supers and has probably started to lay there. If the queen is in the supers the best option is probably to transfer her to the parent colony and hope for the best.

Known causes of failure

Our failure

At the time the artificial swarm was made, the colony was almost certainly on the verge of swarming. Either there was insufficient time for the flying bees to return to the artificial swarm or the weather prevented this happening. With the flying bees still present, the parent colony retained the urge to swarm and did so at the first opportunity (but is was quite a small swarm).

Another known failure

Being nervous about having queen cells and the queen together in the parent colony, one beekeeper destroyed all the queen cells himself during the first manipulation instead of allowing the bees to do this in their own time. This altered the behaviour of the parent colony which swarmed at some time during the next few days.

Appendix 4
Failure of the prime swarm

The practice of wing-clipping the queen has already been discussed in Chapter 2.1. Clipping causes failure of the prime swarm but the beekeeper should be aware that this can occur naturally from time to time.

It happens when the old queen, for reasons that are not usually obvious, is unable or unwilling to fly. She falls to the ground and the prime swarm is aborted in favour of the first virgin to emerge. Often there is little evidence as to what has happened, except that you appear to have a colony that has swarmed (it has no queen and no young brood) but does not seem to have lost any bees. The age of the youngest brood will tell you when this swarm occurred.

There is another possible outcome to this situation about which you need to be aware. A non-flying queen will sometimes crawl under a hive stand or similar refuge and the swarm will try to establish itself in an unsuitable (for them and the beekeeper) place.

The swarm can be recovered from where it has settled and housed in another hive. But, if it has been there for some time before it is noticed, it will have relocated itself to its new position and it will be necessary to remove the hive to a

A swarm that took up residence under an open-mesh floor

completely new location – theoretically at least three miles away but a shorter distance will usually suffice.

The remedial action for the parent colony (the one from which the swarm has issued) is the same in all cases; the queen cells must be reduced to one in order to prevent a cast swarm (see Chapter 2.3, Step 5 for details). Alternatively, the date when the queen cells are due to mature can be estimated in order to do some 'virgin releasing' (see Chapter 2.3, Step 7).

Appendix 5
Bait hives – setting up and management

Bait (or catch) hives are means of capturing swarms; either from your own colonies (your mistakes), from somebody else's colonies (their mistakes) or from feral colonies. Studies of how swarms select nesting sites in the wild and experiments in which swarm boxes of various sizes and designs have been placed in different locations provide useful information of how to set up a bait hive with the best chance of attracting a swarm.

SWARMING BIOLOGY AND CONTROL

What do scout bees look for when selecting a nest site?

Size of cavity

For temperate races of bee the optimum cavity size is about 40 litres. The scout bees have instinctive means of assessing volume by walking in (pacing out) a potential cavity (see Tom Seeley's *Honeybee Democracy*, Chapter 3, page 68).

A Modified National deep box has a volume of about 35 litres, so makes an almost ideal bait hive. The shape of the cavity does not seem to be an important factor but being reasonably dry is, and south-facing is a minor advantage. In hot climates some shade is said to be preferable but that is not a common issue in the UK.

Nucleus boxes are too small to provide an 'ideal home' but they do sometimes attract a swarm which can be something of an embarrassment if it is a large one and all the bees can't get in. If this happens, the swarm will often have second thoughts and abscond.

Entrance hole

This needs to be quite small and an area of 10–15cm^2 is considered to be optimum. Being located somewhere near the bottom of the cavity is also important. When the bees are

An overflowing nucleus

choosing an entrance it has be of a size that will suit all seasons; for thermoregulation in winter and defence when the colony is in a vulnerable condition, eg, re-queening after swarming.

Height above the ground

This is an awkward one – for the beekeeper that is! Ideally a colony would prefer to set up home at a height of 5–6 metres! Obviously it is somewhat difficult to manage hives at this sort of height but it is an advantage (an extra attraction) if the bait hive is located at a height of 2–4 metres – but nothing too heroic. One of our most successful sites is on a barn roof at a height of about 4.5 metres but, for safety reasons when using a ladder, the hive has to be strapped up and lowered on a rope – which can be interesting!

Evidence of previous occupation by bees

The presence of old comb (or rather the smell) is a powerful attractant for a prospecting scout bee but wax foundation is of little interest. Contrary to popular opinion, a free offer of honey is not an inducement – not that it would stay there long before it was robbed out. We have never tried any of the commercial swarm attractants (vials or wipes) or lemongrass oil which is claimed by many to be effective.

Distance from the parent colony

It is well established that a swarm prefers to occupy a new site that is some distance from its previous home. Figures of 50 metres up to 800 metres have been suggested by various studies. This means that a bait hive located within an apiary is less likely to catch a swarm originating in that apiary than if it were some distance away. However, being located in or near an existing apiary is an advantage when attracting incoming swarms which seem to home-in on places where bees are already present.

The ideal bait hive

Making dedicated bait-hive boxes is recommended by some authorities but the use of standard beekeeping equipment is fine – and saves the problem of transfer from the bait hive to a working hive. So, starting from the bottom up, this is what is required in terms of equipment:

- A floor – and an old solid floor is ideal but if a mesh floor is used the catch-tray should be installed.
- An entrance block – the floor should be fitted with a block which gives a fairly narrow entrance with an area of some 10–15cm^2.
- A hive box – as already noted, a standard Modified National deep box is about the right size for a bait hive. An old brood box with a bit of mileage on the clock and the odour of previous occupation is ideal, providing it is reasonably sound. Joining the box to the floor makes for easier handling and fitting a couple of mirror plates (one each side) is a simple solution.

APPENDICES

Standard beekeeping equipment can be used to make a bait hive. The most attractive place for it is 2–4 metres above the ground.

- A cover board – for goodness sake do not forget this because sorting out a swarm that has started to build its nest in the roof is something of a trial. Comb on the cover board is inconvenient but in the roof it is a nightmare!
- A roof – this needs to be sound and waterproof and should have a rock placed on top to prevent it being blown off in what may be quite an exposed position.

What to put in the bait hive

As already noted, evidence of previous occupation by bees is an important attractant and this can be achieved by installing an old, black, smelly brood comb – the sort that would normally be set aside for recycling. However, the beekeeper must ensure that there is no risk that this comb is carrying disease. American foulbrood (AFB) is the main concern because spores of this disease remain viable for up to 40 years. If there has been no AFB in the area for some years and you know the history of the comb the risk should be minimal.

SWARMING BIOLOGY AND CONTROL

But what else (if anything) should you put in a bait hive?

If you are confident that you are going to detect the presence of a swarm on the day of its arrival nothing other than the attractant comb is required. At the other extreme (the belt-and-braces approach), the beekeeper can complete the full complement with frames of foundation. These will not add to the attractiveness of the bait hive and if a swarm does not take up residence in the next few weeks, the foundation will become a bit 'weary' and by the end of the season may be virtually useless – and that is £8–10 down the drain! An intermediate solution is to provide 2–4 frames of foundation – just enough to give a few days' breathing space if the occupation is not detected immediately.

Yet another solution is to install starter strips in some or all of the frames – a sheet of unwired foundation will provide strips for several frames at reduced cost. If you are really clever, you could even wire the frames before fitting the starter strip and then stand a good chance of the bees drawing a perfectly acceptable (reinforced) frame of comb.

Installing starter strips reduces the outlay

APPENDICES

A 'Frankenstein' frame – pieces of comb tied into a frame with wool

What you are trying to avoid is the bees wasting their effort on building comb that cannot be easily incorporated into the hive or brood that cannot be taken full-term. Cutting out and tying wild comb into frames is a tricky job but it may necessary if it contains a significant amount of brood. This has the further disadvantage that the stitched together ('Frankenstein') combs are often spoilt by brace comb.

Signs of impending occupation

There is usually plenty of warning that a bait hive is at least on the shortlist for occupation by a swarm. Increasing interest by scout bees is an indicator – but not a guarantee – that a swarm will arrive and scouting can be so intense at times, with bees coming to and fro, that it is difficult to be sure that a swarm has not already arrived.

If in doubt, go to the bait hive just as it is getting dark; put your ear to the hive wall and listen for the dull roar of bees in residence. Alternatively, give it a sharp tap which will result in a hissing sound if there are occupants. You should also be aware that, if there is competition for a bait hive, some scout bees may bivouac there overnight to deter the competition (a bit like queuing overnight for the Boxing Day sales!).

If a bait hive is being heavily scouted during the daytime and activity suddenly stops, don't be disappointed as this is often an indication that a decision has been taken. The scouts will have returned to the cluster to warm up the swarm in preparation for take-off and this takes about one hour. Wait and watch and you may be privileged to see the swarm arrive.

What to do when the swarm has arrived?

Unless the bait hive is already full of frames, the first task – before any attempt is made to move it – is to make up the full complement with frames of foundation. This should be done carefully; the festoons of bees on the cover board should be shaken into the hive and the frames allowed to 'sink' into the box through the mass of bees – there is no need to force them.

If the bait hive is to be moved, the beekeeper should wait until it is getting dark and all the bees are at home before blocking up the entrance, strapping the hive and moving it.

If the swarm arrived during the afternoon, very few bees will have fully programmed themselves on the new location and the bait hive can be moved directly to its final destination, even if that is only a short distance away. If, however, the swarm arrived earlier in the day, giving the bees more flying time, moving it to a nearby location can be problematic and it may be best to initially take it to a more distant location. Because the flying bees have not yet extended their range very far, this need not be as distant as three miles.

What if the swarm has been in residence for several days?

Now you do have a problem! Before you attempt to move the bait hive to a new location, you must find out whether it is a prime swarm (with an already mated queen) or a cast swarm with a virgin queen which is ready (or actually engaged) in making her mating flights.

The size of the swarm gives some indication but you really need to be certain. The only way to do this is to inspect the colony (outside mating hours, of course) and check for eggs – if there are any they will usually be on the old drawn (attractant) comb. Just to be awkward, some already mated queens may lay eggs on the first night of occupation whilst others may delay for up to five days or more.

I am afraid you just have to be patient because if you move the colony whilst a virgin queen is in the process of mating you may really screw things up – and prevent the colony from re-queening. As soon as you can see eggs, it is safe to move the bait hive but the longer it has been where it is the further away it will be necessary to move it without the risk of losing flying bees.

Feeding the swarm

Regardless of the weather, it is always best to feed a swarm until it has drawn a full set of combs. We always start feeding a swarm immediately it has been given a full complement of frames with foundation and is in its final position. We do not hold off for 48 hours as suggested by many books. A delay in feeding claims to be a means of reducing the transfer of any disease organisms by forcing the wax-makers to exhaust the stores they have brought with them in their honey crops. However, if there is a nectar flow, the colony will start to feed itself anyway (albeit a bit more slowly) – so what's the point?

If disease is a possibility, delayed feeding is a good precaution but otherwise it just delays the drawing of new comb.

The number of combs a swarm will start to draw depends on the size of the swarm. A really large swarm will start to draw a whole box of frames simultaneously, whereas a small cast swarm may limit activity to two or three. When the first batch of combs has been drawn, the colony will not attempt to draw more foundation until what they have already constructed is substantially occupied either by brood or stores.

On one hand, the beekeeper wants to encourage brood production but, on the other, does not want the syrup being fed to be stored in large amounts. The overall aim is to get a full set of combs as soon as possible.

The bees can be encouraged to draw more comb by moving drawn (but not yet laid in) combs to the outside of the box and moving frames of foundation into a position next to the current brood nest. This sort of management must be tempered by common sense depending on the size of the swarm. No amount of feeding and frame moving will enable a small swarm to draw comb quickly and aggressive frame management can impair brood production and thus delay colony development.

To give an example of what can be done with a swarm, we induced a good-sized swarm to draw 22 deep frames in a fortnight by a combination of feeding and frame movement. In this case we were harvesting combs as soon as they were drawn for use in nuclei. This sounds a bit like exploitation but the colony did not seem to suffer as a result.

Conclusions

A swarm is a nice 'freebee' (yes, that is a deliberate pun) and what self-respecting beekeeper does not want that? More seriously though, catching your own swarms can go some way towards saving your honey crop and catching other people's swarms can add to it.

If you catch feral swarms they are a valuable source of genetic diversity. Even though they may have originally come from your hive or somebody else's before establishing a nest in the wild, don't forget that they have survived out there without the help of a beekeeper and been sufficiently successful to be able to swarm. Such colonies have been subject to natural selection and should be regarded as a valuable genetic resource.